9/06 - to RU   NOV 03

DATE DUE

1/10/04			
2/14/04			
MAR 25 '04			
DEC 27 07			
GAYLORD			PRINTED IN U.S.A.

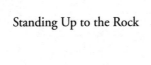

Standing Up to the Rock

Standing Up to the Rock

T. Louise Freeman-Toole

UNIVERSITY OF NEBRASKA PRESS

LINCOLN AND LONDON

© 2001 by the University of Nebraska Press

Manufactured in the United States of America

⊗

Library of Congress Cataloging-in-Publication Data
Freeman-Toole, T. Louise, 1957–
Standing up to the rock / T. Louise Freeman-Toole.
p. cm.
ISBN 0-8032-2011-1 (cl.: alk. paper)
1. Freeman-Toole, T. Louise, 1957– . 2. Women ranchers—Hells
Canyon (Idaho and Or.)—Biography. 3. Ranchers—Hells Canyon
(Idaho and Or.)—Biography. 4. Ranch life—Hells Canyon (Idaho
and Or.) 5. Hells Canyon (Idaho and Or.)—Environmental conditions. 6. Snake River Valley (Wyo.-Wash.)—Environmental
conditions. 7. Hells Canyon (Idaho and Or.)—Biography.
8. Hells Canyon (Idaho and Or.)—Description and travel.
I. Title.
F752.H44 F74 2001
979.5′7—dc21
[B]
2001025341

To the memory of my parents, Pete and Peggy Freeman

Contents

Illustrations

Acknowledgments

I would like to thank Dr. Doug Hesse, David Foster Wallace, and Dr. Rodger L. Tarr for overseeing the beginnings of this manuscript as a master's thesis at Illinois State University; Mary Clearman Blew for her unflagging support through the years; Kim Barnes and Rick Spence for their advice on publishing; and Terry Tempest Williams for allowing me to introduce her to "Ram's Head Rock." Special thanks to Clark Whitehorn, Gary Holthaus, Charlene Porsild, and Ron Landeck.

For reading early versions of the manuscript and offering insightful comments, I would like to thank Keith Petersen, Karen Young, Mary Dye, Sue Hovey, Paula Coomer, and Dale Gentry.

Thanks also go to Bill Rybarczyk and Ed Schriever at the Idaho Department of Fish and Game, who provided information on Craig Mountain and the salmon recovery program; Mark Elsbree of the North Idaho Field Office of the Nature Conservancy, who answered my questions about Garden Creek Preserve and logging on Craig Mountain; and the late Ipsusnute (Jesse Green), who told me about Nez Perce fishing on the Snake River.

I am indebted to Helen Beard and Lee Earl for offering their perspectives on local history; Vada Repta and Eldora Fitting for sharing their childhood memories of the Snake River Canyon; and Jedd Gould for giving me an update on Square Island.

The Charles C. Redd Center for Western Studies at Brigham Young University contributed much-appreciated financial support. "Waiting for Coyote" appeared in a different form in *Written on Water: Essays on Idaho Rivers*, ed. Mary Clearman Blew (Moscow: University of Idaho Press, 2001).

I appreciate the able assistance of staff at the following institutions and agencies: Nez Perce County Historical Society Luna House Museum, Lewiston, Idaho; Asotin Museum, Asotin, Washington; Washington State

University Manuscripts, Archives, and Special Collections; University of Idaho Special Collections and Archives; University of Idaho Photographic Services; Whitman County Parks and Recreation; Ketchikan–Misty Fiords Ranger District; and the State of Alaska Department of Natural Resources, Division of Lands, Juneau office.

Thanks to Richard Storch for loaning photographs from his collection; Beth Rimmelspacher for her artistic vision of the Snake River; and Veronica Lassen, Adrien Sinclaire, Richie McPeek, and others who accompanied us to the ranch on various occasions.

The following people encouraged and sustained me in various ways throughout the writing of this book: my sisters and brothers, Lisa, Leslie, Kerry, Julie, Taylor, and Ingrid; my Redondo friends, Kathy Lewis, Anita Van Meter, and Darla Jasmine; my Illinois relatives, Dorothea Babb and Judy Kay Inman; and my godfather, Jerry Luethi.

I am truly grateful to the woman I call Bertie for introducing me to the ranch and to "Dooley and Liz Burns" for letting me into their lives long enough to write this book. Thanks to my son Ambrose for being my nature companion both on and off the ranch. Special thanks also go to two people who do not appear much in this book, but who are a very important part of my life: my son Emlyn for his steady good cheer and for our many conversations about writing; and my husband, R., for everything, but especially for understanding that "the ranch happened to you."

ONE

Heading North

In our wedding vows, my husband and I pledged to live together in a "green and peaceful place." For ten years Santa Cruz, California, had fulfilled these requirements. We lived in a cabin surrounded by redwoods and later moved into town just a few blocks from the cliffs overlooking Monterey Bay. Santa Cruz was a haven for hippies, artists, and street people. In the 1970s it was still uncrowded and cheap enough for us to survive on part-time jobs while we spent the majority of our time on creative projects. I worked in my darkroom while my husband composed music at the keyboard; together we wrote musicals and put on concerts. But by the mid-1980s San Jose had turned into something called Silicon Valley, and computer geeks discovered that a house at the beach was well worth the daily commute through the Santa Cruz Mountains. In a town where people usually walked, took the bus, or rode bikes, traffic had never been a problem; suddenly there were cars everywhere. The price of housing shot up and we realized we would never be able to afford a home of our own. The sometimes sweetly addled street people with guitars and flowing skirts were being displaced by an increasing number of people with serious mental illnesses.

Aspects of Santa Cruz that hadn't bothered me before assumed a new importance once we had children. With a baby in a frontpack and a two-year-old toddling ahead of me, it suddenly mattered that the downtown sidewalks were covered with spittle. The drug transactions in the park, to which we had always turned a blind eye, were not so easy to ignore with a child playing on the jungle gym just a few feet away. A new jail built to house the growing number of petty criminals turned into a municipal embarrassment when inmates started escaping with alarming frequency. They often came through our neighborhood, heading for San Francisco, seventy-five miles away. One high-speed chase in a stolen car ended with the escapee smashing into a friend's van parked in front of our house. The next time a prisoner escaped, three-year-old Emlyn looked out the kitchen window and asked, "Mommy, why is there a policeman in our backyard?" I looked outside and saw a cop with his gun drawn crouched on the shed roof. This was not our definition of "peaceful." Santa Cruz had been a good place to be starving artists, but it was no place to raise children.

The summer my mother died and our son Ambrose was born, we took stock of our lives and decided it was time to move. A year later we had saved up enough money and made a tentative decision where we would go. I wanted our kids to grow up knowing their aunts and uncles and cousins, but we couldn't bear to move back to Los Angeles. We had already checked out Colorado, where one of my sisters lived, but Boulder seemed like Santa Cruz with snow. My younger sister and her husband had recently moved to Idaho. Ingrid urged us to check out the Palouse—an agricultural area along the Idaho-Washington border. She described hills that looked like sand dunes covered with golden wheat. She talked about how happy they were to have left Santa Barbara—a town, like Santa Cruz, that was increasingly split between the beautiful people and the homeless. At a loss to describe exactly what was so appealing about the town of Moscow, she finally blurted out, "You can smell things growing here!"

We decided to take a trip up to Idaho and stay with my sister and her husband for a few weeks while we looked around the area. We hadn't been on the Palouse but a week before we decided this was it—the green and peaceful place in which to raise our boys. Richard rented a truck, drove to Santa Cruz, packed our belongings, and said good-bye to California. The boys and I, meanwhile, settled into a rented house on a shady, maple-lined avenue with the Beverly Cleary-esque name of Spotswood Street.

After living in Moscow for a year, we bought a house across the border in Pullman, Washington, eight miles from Moscow. We, too, were taken with the Palouse. The small towns were safe, the people friendly. You could walk anywhere you needed to go. Standing on Main Street, you could see combines working the hills outside of town. Every day as I walked to my part-time job at the public library, I witnessed small scenes I thought belonged to some vanished America: a bus driver leaning out his window to take a cup from a little girl selling lemonade on the corner; a boy sitting on a kitchen chair on the front lawn getting his hair cut; young men in ROTC uniforms posing for a group portrait in a city park. My husband and I were giddy with delight. It was like being able to take our children back to the time we had grown up in—a safer, slower, and kindlier world. There hadn't been a murder in twenty years. Front page news might be the bear that wandered into town or the crop damage caused by some callous youth driving through a farmer's wheat field.

In Santa Cruz, theft was commonplace, and we had things stolen even at church. Yet on the Palouse, women left their purses in the grocery cart while they shopped. Locals locked their doors only if they were going out of town for a few days, and not even then if the neighbor needed to come in to feed the cat. They took an equally casual attitude toward their children. When Emlyn disappeared after church one day in Pullman, I ran, frantic, through the halls looking for him. A woman stopped to pat me on the arm and say, "Relax, dear, it's pretty hard to lose a child in this town."

We were so dazed by the novelty of it all—the county fairs (women still made jam and canned green beans!), autumn leaves, Fourth of July picnics—that it took some time for us to realize that all these things added up to something fundamentally different from what we had known in California. A few months after we arrived, my sister and I were driving up the steep grade that led from the Lewis-Clark Valley to the Palouse. We were startled to see a cow jogging heavily in the downhill lane on the other side of a cement barrier. She looked exhausted—her tongue hanging out, the whites of her eyes showing as she lurched out of the way of a truck barreling down the grade, horn blaring. She had probably blundered onto the highway through a break in the barbed wire. Now she was hemmed in on both sides and still had several miles to go to the top of the grade.

"We should do something. Somebody's going to come around a curve and hit her," I said. Then, looking ahead for one of the red emergency phones I used to see along California freeways, I said, "Let's find a call box." Ingrid broke into laughter. "They don't have call boxes in Idaho. Sis, you're not in California anymore."

In Idaho, I learned, you don't call for help. In Santa Cruz, there was no shame in asking: people asked for spare change, food stamps, a place to crash. Here, help may be offered by friends, neighbors, and strangers, but you don't ask for it. The emphasis is on hard work and self-reliance—standing on your own two feet, overcoming your fears, and facing up to the situation—on what old-timers used to call "standing up to the rock."

Two years after we arrived in Idaho, I met Liz Burns, who was to teach me much about the Idaho way of life. She is a woman in her early forties, who, with the help of her elderly father, assorted hired hands, and three well-trained cow dogs, runs a cattle ranch about sixty-five miles south of Pullman. The Burns Ranch lies along the last free-flowing section of the Snake River—the 110-mile stretch between Hells Canyon Dam and the small town of Asotin. The ranch was homesteaded by Liz's great-grandparents in 1894 and is the only one in the area still owned and operated by the descendants of the original homesteaders.

Liz's father, Dooley, now in his eighties, lives in the back of the old schoolhouse at Burns Ferry, where he attended school through the eighth grade. He plants a large garden every year, makes mulberry wine, and keeps the equipment in good repair. Still remarkably fit, he rows easily back and forth across the river and can hike for miles over the hills hunting elk. Dooley spends much of his time in the small town of Asotin, where he is restoring an old steamboat permanently docked near the town park.

Liz lives in a large old house on the other side of the orchard from the schoolhouse. Liz was away at college, pursuing a double major in psychology and religious studies, when her mother died. She returned home to help with the cattle operation, which she now runs with the help of various friends and hired hands who come from as far away as Montana and Alaska and stay for anywhere from a few weeks to a few months. The most regular of these ranch hands—a woman named Bertie—introduced us to the ranch.

Every few weeks, I would drive through wheat and lentil fields, descend the two-thousand-foot grade into Lewiston, Idaho, cross the Clearwater River, then cross the Snake River back into Washington again and follow the river south. The paved road changed to gravel littered with football-sized rocks dislodged from the canyon wall above. I would see bighorn sheep on the rimrocks across the river or a solitary heron standing in the shallows. Two hours after leaving home, I would get into a rowboat left chained to the rocks at Burns Ferry, row myself across to the Idaho side of the river, and climb the steep sandy bank to the house alongside Thornbush Creek.

My husband and children loved the Burns Ranch but were content to visit once or twice a year. Although they enjoyed the horseback rides and other amusements, the normal workaday life of the ranch held little appeal for them. I was content, on the other hand, to hang out and do whatever was on the agenda for the day, whether it was weeding the garden or fixing fences. For several years, until Ambrose grew old enough and started begging to come with me, I usually went to the ranch by myself. As a homeschooling mom who spent many hours alone with my kids, I appreciated these occasional breaks. And Richard, who saw little of the children during the week, was always happy to have them to himself for a day or two.

I came to the Burns Ranch in spring to help round up the new calves, and in summer to kayak the river and pick apricots. I returned in autumn when the tourists leave the littered beaches to the locals, and again in winter when the only human figure on the river is an occasional solitary steelhead fisherman. Most of all, I came to go to the rock.

From Burns Ferry, we would drive several miles along the river on a rock-strewn track clinging to the side of the steep bluff, arriving at Ram's Head Rock in a flurry of dust and dogs barking. The dogs, who had been running along behind, would go down to the water to drink next to Ram's Head Rock—a jumble of ancient stones so unyielding that the river has had to detour around it.

Across the river, the west wall of the canyon rises steeply to the Anatone Prairie, an isolated area with more hawks than people. The east side of the canyon ascends in a series of bluffs and side canyons up to fir-covered Craig Mountain and onto the Camas Prairie, where the Nez Perce used to dig for camas roots. Upriver, to the south, lie the further reaches of Hells Canyon, where tourists drive to Hat Point to throw rocks into the country's deepest chasm. Beyond the prairie to the west are the peaks of Oregon's Wallowas, from which Chief Joseph and his band fled northeast through Montana pursued by the U.S. Cavalry. Downriver, at the confluence of the Snake and the Clearwater River, is Lewiston, with its fragrant orchards and stinking paper mill.

The first time I stood at Ram's Head Rock, looking out over the broad field covered with yellow flowers, the curving river, and the hills of bleached grass, I felt a childish sense of envy. Liz had grown up at Ram's Head. She climbed on the rocks, swam in the river, rode horses along the beach. She had all this space to grow up in. I thought of my own suburban childhood— the sprinklers, the noisy neighborhood games, the bikes and skates, the hot afternoons at the beach, and although those memories were precious to me, I would have gladly traded it for this. If I had been eight years old, I would have run across the pasture with my arms open wide, embracing it all.

My son Ambrose responded the same way. After that first visit to Ram's Head, we stood in the bed of the pickup as Liz drove the rough road back to Burns Ferry. Ambrose clung to the bars of the stock rack, knees flexed to take every bounce, mouth open like a dog tasting the wind. "Mom," he shouted above the roar of the truck, "How do you sing hallelujah?"

TWO

The Cow Business

The newborn calf is lying on a patch of brilliant green moss among the rocks at the bottom of a narrow ravine. "There it is! It's alive!" Ambrose yells. At the sound of his voice, the calf scrambles to its feet just as Bertie and I fight our way out of a hackberry bush several yards away. The calf appears unharmed by the fall from the eight-foot rock ledge above. Ambrose bends over and puts his hands on his knees, catching his breath from the uphill climb, his eyes never leaving the little brown calf. There's a long red scratch on the side of my son's face, and his pants are wet where he slipped on a moss-covered rock and fell into one of the small springs that well out of the ground every few yards.

The calf licks its nose and stares back at Ambrose. The twisted rootlike end of the umbilical cord hanging from the calf's belly looks like something in the window of a Chinese herb shop. He's been separated from his mother since morning; he must be hungry. If we hadn't found him by nightfall, he would have made an easy meal for the coyote we saw disappearing over the ridge with a backward glance over his shoulder. My son is a sucker for lost little things. But the calf is not glad to see us. As Ambrose approaches cautiously, the calf blows a string of mucus from one nostril and stumbles over a rock trying to get away.

We met Bertie the year after we moved to Idaho, when my husband became the new conductor of a community orchestra in Lewiston. Over the next few years, it became a family affair; I played viola, and Emlyn played second violin (until age nine, at which time he switched to oboe). Ambrose came with us to rehearsals and lay on the floor, drawing.

Every autumn, the talk at rehearsal ranged from music to hunting to more hunting. One night I missed rehearsal because Ambrose was sick. The next week, a flute player said, "You should've been here last week.

Reeve got his elk. He had it out in the parking lot, so we all took a break and went out to see it."

Reeve, a local cop who plays second violin, said, "Yeah, we skinned it in the back of my pickup." He smiled widely at my horrified look. He shrugged and pointed at Bertie, "Hey, she wanted the hide." Bertie, the french horn player, was by far the best musician in the orchestra, but she was so quiet and unassuming that I hadn't really noticed her. Now I looked at her with new interest.

I discovered that she worked at a ranch down in the Snake River Canyon. Every week after orchestra rehearsal, she drove for an hour on gravel roads and then rowed across the river in the dark to the ranch house. The only other people on the Burns Ranch were old man Dooley and his daughter, Liz. Dooley wasn't quick enough on his feet to work with the cattle anymore, and Bertie had been hired to take his place. To fill the long evenings on the isolated ranch, Bertie had taught herself to tan hides using deer brains. She used the elk and deer hides to make everything from drums to rawhide whips.

I got to know Bertie over the next few months. On her way in to rehearsal, she always stopped to say something to Ambrose or to admire the picture he was drawing. Sometimes she showed him an arrowhead or a pretty rock she had found on the beach. Bertie invited us out to the ranch several times that winter and the following spring. We played music together, hiked up the side canyons, and went kayaking on the river. But it was not until that summer that I had anything to do with the main business on the ranch: cows.

One night as I sat tuning my viola before rehearsal, Bertie leaned over and said, "Liz and I going to start roundup tomorrow."

"Just the two of you?"

"Unless you want to come help," she said only half-seriously.

"I wouldn't be much help. Remember, I'm no good on a horse."

"Oh yeah," she laughed, recalling a recent visit in which the horse I was riding made a beeline for an apricot tree down by the river. Gramps, ignoring my kicks, commands, and desperate pleading, ran under the overhanging branches of the tree to get at the ripe apricots. I managed to stay in the saddle but got so scratched up I swore it was the last time I was ever going to get on a horse.

"Well, that's OK. Liz and I are going to ride up into the hills tomorrow and bring the cows down. By this weekend, they'll be down by the barn and then we do everything on foot." Bertie was rather vague about what exactly we would be doing, but she assured me I would be useful. "Bring

Ambrose along," she said, looking over at him lying on the floor drawing stick figures of knights in battle.

When Ambrose was three years old, he had insisted on being called "Cowboy Bob." He roamed the yard in boots and jeans, a kerchief around his neck and a cowboy hat pushed back on his head, blowing on a harmonica. He has long since outgrown that phase, but he loves going to the ranch. I had given up trying to get his older brother and his father to come to the ranch. Emlyn has little use for nature. On the infrequent occasions when he comes to the ranch, he spends most of the time indoors, playing computer games or reading fantasy novels. If we drag him along on a hike, he spends the entire time complaining. My husband likes to visit Ram's Head, but the daily life on a cattle ranch holds little appeal for him. He dislikes animals and bugs, both of which are found in ample supply at the ranch.

That day in early June when Ambrose and I came for roundup, I saw Liz waiting for us on the other side of the river—a short, compact figure wearing rip-proof coveralls and massive boots. At that time, I hardly knew Liz. She was wary of newcomers and tolerated us because we were Bertie's guests. I was a little afraid of her. Bertie landed the boat on the beach and Ambrose jumped out to play with the cow dogs. When I stepped onto the sand, Liz looked down at my soft suede shoes and snorted, "That's all you got on your feet?" I nodded sheepishly. "You'll have to wear some of my boots."

Up at the house, Liz handed me a pair of huge iron-hard boots and some chaps with a complicated-looking set of buckles.

I hefted the heavy leather chaps. "I'm really supposed to wear these?"

"Did you bring a change of clothes?"

"No."

"Well, then you oughta wear them. Be a long drive home covered in cow shit."

She also gave me a pair of stiff leather gloves and a red cap that said "Whipple's Feed and Wild Game Cutting."

Six-year-old Ambrose took one look at my outfit and said, "Mom, you look silly."

I felt silly. I was just glad they hadn't clapped a cowboy hat on my head. But one of the surprises of coming to the ranch was finding out that ranchers don't wear cowboy hats: it's the people in town who wear them. Ranchers and farmers wear caps from the local feed stores. I liked Liz's cap better. It was tan and said "Lewiston Grain Growers."

Cowboy Bob, age three. Photo by the author.

We walked out to the barn. I discovered that to walk in the heavy chaps and boots, I had to swing my leg around to the side with each step. I got into the rhythm of it and pretty soon I was clanking and rolling over the rough ground like a sailor walking the deck in a high sea. Dooley came by to find out what the plan was for the day. He glanced at my ridiculous getup. "Playin' cowboy, huh?"

"No, Pop, she's just interested in the cow business."

I was grateful that Liz understood why I was there. It wasn't because I had some romantic vision of myself sitting tall in the saddle above a windswept vista. I was just fascinated by the thought of these two women having the know-how and the guts to do this roundup. My natural inclination would have been to come and observe, but I understood that on a working ranch, there is always a lot to do and everybody is expected to pitch in.

Liz explained that we would be moving a group of cows from the field on the other side of the creek up and into the corral by the barn. Liz told Ambrose to stay at the barn so he would be out of the way. "Just don't slide on the hay. It breaks up the bales and makes a big mess," Liz warned him. Ambrose looked disappointed until Bertie told him there was a mama cat with kittens in the barn. They were pretty wild, but if he was quiet they might come out where he could see them.

Out in the field, Liz told me, "You have the easy job. All you've got to do is stand over there and keep the cows out of Pop's potato patch. Bertie and I will get them moving across the creek to the barn."

Sounded simple enough. Bertie gave me a swat—a flexible rod about four feet long, like a longer version of a riding crop. "If they come your way, just wave this and yell."

I took up my position in front of the newly planted potato patch and looked across the field at the cows and calves. They were standing as far from us as possible, all heads—big and little—turned in our direction. Liz and Bertie moved to opposite sides of the field and started closing in on the cattle, whooping and waving their swats. Though the cows stayed in a tight-knit pack, they didn't turn like a flock of birds. They ran this way and that like a bunch of Keystone Cops, wheeling around and bumping into each other, desperate to get away from the loud figures coming at them. I realized then that they were not barnyard cows like the ones I often saw on the Palouse. These were range cattle: they had been up in the hills for months, seldom seeing a human being on horseback, let alone one on foot. As the two women funneled them closer to the creek crossing, the cows panicked and turned, thundering in my direction. There were sixty or seventy of them charging toward me, looking both scared and mad as hell. My first impulse was to run for the river, but I had to defend Pop's potatoes.

I trundled back and forth, chaps flapping, waving the swat and yelling as loud as I could. But I didn't know exactly what to yell. Bertie and Liz used a wordless, warlike cry, but I couldn't do it right. I ended up running back and forth in front of the potato patch, shouting, "Hey! Hey!" sounding like a ninety-eight-pound weakling getting sand kicked in his face. No wonder the cows ignored me.

When I saw they weren't going to turn aside, I stopped running and raised my swat in the air. Like the parting of the Red Sea, the cattle separated in a neat V around me and trampled mightily through the potato patch. Slowed down by the freshly plowed earth, they jogged heavily around, each step sending some poor one-eyed potato down too far to germinate. I finally got the cows out of the patch, but they scattered across the field so that Liz and Bertie had to start the whole process over again. Just then I noticed Dooley standing by the creek watching. I felt my face go red. Some cowboy.

Dooley had a knack for showing up at the worst possible times. Months later, after Bertie and Liz grew tired of ferrying me back and forth across the river, Liz announced it was time I learned to row myself. I still didn't have good control over the oars and had to row a long way upriver before I started across to make sure I arrived, drifting and struggling, at the beach in front of the house. One time I couldn't seem to coordinate my oar strokes, and the boat started going in circles. Ambrose said, "Mom, the beach is that way." Just then I saw Dooley's truck slow down on the road above us. He must have been wondering who in hell was going in circles in the middle of the river in his boat. He stared for a minute, then drove on to the garage. We landed downriver on a rocky beach and had to walk the boat back upriver to the anchor chain.

Once we had gotten the cows across the creek, moving them into the corral was easy: there was nowhere else for them to go. Ambrose walked carefully out of the barn, his hands cupped in front of him. "Look what I found!" he said. It was a mouse the size of his little finger. Its eyes weren't even open. I knew what he was going to say next.

"Can I keep him?"

"Well, honey, he looks pretty little."

Bertie bent over Ambrose's hands and stroked the mouse. "The barn cats probably got the rest of the litter. We can try feeding him with an eyedropper." She took Ambrose back to the house to nurse his new pet.

I was glad that Ambrose was gone. He had already said he didn't want to watch us brand the calves, and I didn't blame him. The electric branding iron was warming up on the fence, sending heat waves into the air. It looked more lethal than the old-fashioned poker they used to put in the fire.

With some prodding, Bertie explained that we were sorting the cattle into cows, calves, heifers, and steers. The steers (castrated males) were put into a separate pen, where they would stay until they were trucked, a few at a time, over the mountain to the auction house in Lewiston. The heifers (young females without calves) were given a vaccination and returned to the hills. All the calves had to be branded and the male calves castrated. The cows (females with calves) were vaccinated and then reunited with their calves.

This was the first time Liz had seen the cattle close up since the previous autumn. She looked them over carefully, commenting to Bertie that it was a good crop of calves. She noted which cows hadn't calved and would have to be sent to auction.

The two women calmly and casually discussed what needed to be done. Perhaps it wasn't as complicated and dangerous as it seemed. I couldn't imagine walking into the corral full of spooked cattle all rolling their eyes and tossing their horns, yet a minute later I found myself on the other side of the gate with Bertie, face to face with the cows. They were heaving themselves from one side of the corral to another, hooves thrashing and slipping in the mud. They climbed over each other and sideswiped the fence trying to get away from us.

Liz called out, "Get me that little heifer over in the corner." I gripped my swat and watched Bertie to see what I was supposed to do. She waded into the seething mass of cows and whacked a brown and white heifer on the rump. The heifer leaped toward me, and I cracked the swat in the air, making a sharp snapping sound. She turned and plunged toward the gate. Liz swung it open, the heifer ran through, and Liz quickly slammed the gate, yelling and banging it against the cows trying to escape along with the heifer. Only the heifer hadn't escaped: she had been released into a chute leading to a barred enclosure where she was to be vaccinated. Once the heifer stepped into the enclosure, all three of us worked ropes and levers, simultaneously lowering a barrier behind her, clamping her head in a metal yoke, and pressing the bars against her body to hold her still. Liz filled a syringe the size of a large carrot and plunged the needle into the heifer's flank. Then we worked the ropes and levers in reverse, releasing her and opening the gate to another corral. We went through this routine for each of the heifers. When the more skittish ones wouldn't walk down the chute,

Bertie had to climb in and encourage them along with her swat. One heifer released a load of manure just as Bertie got within close range, covering her from belt buckle to boots.

Then we began working with the calves. As we moved around the corral, the cows maneuvered to keep their babies behind them. The calves leaped about like baby goats. When one of them jackknifed away from me for the tenth time, I looked over helplessly at Liz. She was leaning on the gate, grinning as she watched Bertie and me run around and around the corral. "Those calves are just full of green grass. It makes them goosey," she yelled over the bawling of the cows. Finally Bertie managed to get a calf into the branding chute. We clamped the calf between the bars and Liz pulled a lever, swinging the enclosure over until the calf lay on its side about three feet off the ground. Bertie took a green plastic tag out of a bucket, wrote a number on it with a permanent marker, and recorded it in a notebook. Liz handed me a large staple gun and showed me how to position a tag on the calf's ear. I held the tag against the thick, furry ear and pressed the stapler. The plastic staple went in with a bit of resistance, like sticking the muzzle of a potato pellet gun into a raw potato. The calf flicked its ear, no more bothered than if it had been bitten by a horsefly.

Liz offered to let me brand the calf, but I declined. I looked away from the branding iron and stared into the wild blue eye of the calf. As the iron sizzled against her hide, the calf let out a loud bawl. Her tongue jutted out and foam dripped from her mouth to the ground. From the corral, her mother answered with a desperate bellow and heaved herself against the fence. Liz wielded the branding iron quickly and neatly, burning Liz's brand (her initials, L and B, stuck back to back) into the calf's side. She explained that you had to do it with just the right amount of pressure and for just the right length of time—otherwise the mark would be too light or the burn too deep.

We released the calf, and it staggered off and stood by the fence, sides heaving. It appeared to be shaken but not severely hurt. I too was shaken, by the cow and calf's distress at being separated, the smell of burning hair and flesh, the sound of an animal in pain. I was relieved there were no male calves in this batch so that I was spared the sight of a castration. Liz said, "At least I castrate them with a sterile scalpel. Pop used to do it with his pocketknife. We lost a lot of calves to infection."

By the end of the afternoon, my feet were sore from slipping around inside Liz's boots. I was incredibly tired and sick of the sound of bawling cows. Bertie's clothes were stiff and stinking; she was eager to get back to

the house to change. There were streaks of mud on Liz's face and her hair
had escaped from her braid in tendrils around her face, but she seemed
happy, apparently satisfied with our work for the day.

When we got back to the house, Ambrose was watching TV with the
mouse in a little box on his lap. After an early dinner, Liz rowed us back
across the river. Ambrose jumped out of the boat and clambered up the hill
to the car. Liz said, "It went pretty well today. Nobody got hurt."

I thought I had just been a wimp about how dangerous everything
had seemed that day. "Does someone usually get hurt?" I asked, not really
wanting to know the answer.

"More times than not. Stepped on, usually. That's why you need good
boots."

Next time I would come prepared with sturdier boots and extra clothes.
The long day had been filled with mud, blood, manure, and unending
noise, yet I already assumed I would be coming back to do it again. My
body was tired and my mind was emptied out. Rarely had I spent a day
so utterly absorbed in what was going on around me. The concentration
required to dodge cows, open and shut gates at precisely the right time,
and work together smoothly with the other women had been like a moving
meditation. I began to understand Liz's curious calm. This was her life every
day.

I stepped onto a rock and leaned against the prow of the boat to shove
it out into the current. Liz said, "I got kicked in the face by a horse, back
in high school." She bent forward over the oars and suddenly popped her
front teeth out at me. Startled, I slipped off the rock into the water. She
laughed out loud. "Took out my front teeth. It was a real mess. Had to
have surgery on my gums. Couldn't eat for weeks."

With a strong backward stroke, Liz pulled the boat away from the bank.
Resting the oars in her lap, she cupped her hands around her mouth and
called out, "You gotta be ca-a-a-areful out here." Her words carried across
the water and echoed against the canyon wall. Above me, Ambrose stood
on the road waving good-bye.

Liz and Bertie have been walking for miles every day this month looking
for cows having trouble giving birth. I was watching Ambrose wade in the
river when Bertie showed up and asked for help in finding a lost calf. Earlier
this morning Bertie had come around a rock outcropping and startled a
cow nursing her newborn calf. The cow took off in one direction and the
calf, too young to realize it ought to follow its mother, took off in the other
and tumbled off a cliff. Bertie peered into the ravine to see if the calf was

all right, but rocky overhangs and brambles made it impossible to see to the bottom.

"You go," Liz said to the air in my general vicinity. "I need to check on a couple of pregnant cows up there." She nodded toward several cows grazing above one of the draws. I expressed my amazement that she could tell the cows were pregnant from so far away. She shrugged and mumbled something about the way they walk. Bertie climbed out of the truck with a coiled lead rope, and Ambrose and I followed her up the hill.

Here, I thought, is a job I can handle: walking the grassy slopes between the rimrocks, doing nothing but keeping my eyes open. Ambrose, too, can hike for hours, and he's got sharp eyes. I'm not surprised that he was the first to spot the lost calf. And looking for calves is a good excuse to be out here in the spring when everything is green and blooming. For most of the year, the bluffs along the Snake River are buff brown. The tufts of grass between the rocks are dry and blond as hay. The monotone of the landscape makes anything with color or movement as noticeable as a waving flag. The occasional car coming down the road is visible for miles, spiraling a cloud of dust behind it. The comings and goings of animals are easy to see. Bighorn sheep scramble up the rock outcroppings on Devil's Rim. A bald eagle returns to the same lone pine she nested in last year. In the willows along the river, orioles stitch hanging nests out of dry grass and tangled skeins of fishing line. Deer bound out of the truck's way. I can almost always see a hawk somewhere in the sky.

In spring, the new grass comes in and the near-translucent green goes on and on up thousands of feet to the top of the canyon. The contrast of the delicacy of the color and the massiveness of its scale is disorienting, throwing off all sense of proportion. I feel huge and light on my feet striding across the hillside, stepping over tiny spring flowers like Lilliputian trees. But looking up, I see the opaque blue sky resting on the rim of the canyon like the painted ceiling of a diorama, and suddenly I feel small again.

Together, the three of us corner the calf against a rock wall and slip a rope around his neck. With Ambrose tugging on the rope and the two adults boosting the calf from behind, we get him out of the ravine. Two of the cow dogs have followed us up the hill. They discover the afterbirth in the grass not far from where the calf fell from the ledge. They both grab hold of it and tug. It stretches between them like a strand of pizza cheese as they gulp it down. Ambrose tries to lead the calf but discovers that, unlike a dog that will walk on a leash, this baby has no intention of going anywhere with him. We have to work together to get the calf down the mountain:

one hauling on the rope, one pushing, and one heading him off when he tries to make a break for it.

We stop to rest when we get to the top of a steep bluff. Ambrose runs back and forth picking flowers and jumping off rocks. I sit down in the grass to pick burrs out of my socks. Bertie says, for the ninth or tenth time, how glad she is that the calf isn't hurt. I'm just glad we don't have to carry him out; even as a newborn, he probably weighs eighty pounds. I know Bertie is relieved she didn't do the wrong thing by leaving the calf and coming back for it later.

Bertie has been working on the ranch for ten years, but she doesn't have the advantage that Liz does of having grown up on a ranch. She's from San Francisco, although she seldom mentions the city. She did tell me that as a child she was afraid of everything, especially people. She spent much time walking in a wooded park near her home and reading nature books such as *Rascal* and *Black Beauty* and the works of Ernest Thomas Seton. She loved horses and hoped to own one some day.

In some ways Bertie is like other ranch hands who used to drift into the canyon years ago when there was more seasonal work available on the ranches. They came from distant parts and didn't talk much about their past. They did some haying and irrigating and helped with roundup. In winter they might work as caretaker of a remote lodge or ranch. In summer they did a little placer mining and raised a garden on their own bit of land. When there's not much to do at the Burns Ranch, Bertie also works various jobs in the canyon and up on the prairie—driving combines, "slapping sandwiches" at a resort upriver, house-sitting. She is certified to teach school, but facing a classroom of children is as terrifying to her as facing that herd of cattle at roundup had been to me.

Like many of the old-time drifters who wound up in the canyon, there is a bit of the loner, the oddball about Bertie. She's always developing some new system for picking the winning lottery numbers, such as watching to see which numbered strips of paper are taken up by the wind and whirled like maple seeds across the yard. She tunes into a nightly radio show on unexplained phenomena and is a firm believer in everything from crop circles to UFO abductions. She plays mandolin in a popular bluegrass band. Her room at the back of house is as spare as an army barracks. At the foot of her bed is a duffel bag packed and ready to go for her next band tour. A cowboy hat hangs on a nail in the wall next to a poster of Mozart. In the corner is a small table with two things on it: a jewelry box and a BB gun.

Bertie was not the only hired hand on the Burns Ranch. Through the years, they have also hired an assortment of equally interesting characters, including Dermott, a horsebreaker from British Columbia; Jimmy C., a former beatnik poet; and a massage therapist named Willow, who had grown up on a ranch in Montana.

There has always been room in the canyon for eccentrics and hermits. Back in the 1930s, there lived in the canyon a woman known only as "the wheelbarrow lady." She appeared one day on the trail into Hells Canyon, burdened by two enormous bundles. She would carry one bundle a ways down the road, then return for the other. She made such slow progress that a woman at one of the ranches felt sorry for her and gave her a wheelbarrow. She stayed in the area for several years, eking out a living along one of the creeks, and then one day she trundled her goods out of the canyon without so much as a good-bye.

There was a hermit who spent his time building stone walls up and down the length of his property, sometimes carrying a rock around for days until he found the right place for it. Another man, who worked as a sheepherder and occasional boat hand, was known only as Silver Buckle. Rumor had it that he ran into trouble with the Mounties up in Canada and came to the canyon to hide out.

I ask Bertie what brought her to Idaho. With a little prodding, she tells me.

"It was a guy, what else?" she laughs. When that didn't work out, she entered college. "I took one of those aptitude tests, and it said I'd be happiest working on a ranch. I liked the idea, but it didn't sound like a real job. Then I met Liz in one of my classes and she said she needed help on the ranch. I came out here, saw this place, and . . ." her voice trails off and she just shrugs and smiles. "Where else could I live like this?" she asks, the sweep of her arm taking in everything around us—the green hills, the river below, even the little troublesome calf. She looks utterly content, sitting cross-legged, fiddling absentmindedly with a piece of grass. She holds the blade of grass between her thumb and blows on it. A reedy whistle sounds loud in the quiet, and Ambrose, standing on a outcropping, turns and waves.

Bertie continues, "I made a lot of mistakes when I first starting working out here. I got lost a lot. Once I threw a saddle onto a horse with Liz standing on the other side, and the stirrup broke her nose. You should've heard the way she swore at me. One time I was by myself when a cow gave birth to a stillborn calf. I didn't know what to do. It was dead, right? Well, Liz got really mad at me for not trying to save it. You're supposed to scoop the mucus out of their mouth and then blow into their nose. But I didn't know that back then."

Now Bertie tries to take action of some sort if Liz isn't around, although it doesn't always turn out to be the right thing to do. "The other day I was out by myself and I saw a cow with a dead calf sticking halfway out of her." She tried to pull the calf out by its front legs. Then she tried to stick her arm up inside the cow to untangle the back legs, but it was wedged too tight. She finally resorted to tying a rope from the calf's legs to the bumper of the truck and backing up.

"Turns out that was the wrong thing to do. The calf came out, all right, but it dislocated the cow's hip." Liz was very upset and told her, "You have to know when to give up and call the vet." With her customary practical focus, Bertie was able to salvage something from the situation. "I skinned the calf and I'm going to make chaps out of it."

Since that first disastrous day in the potato patch, I've worked with the cows whenever I can. I've never developed Liz's affection for them, or Bertie's patience. But I've learned something essential about cows: they are incredibly stupid. Every year the Burnses lose at least one cow that gets itself stuck on a precarious spot on the face of the cliff. If the cow is close to the valley floor, Liz can get it down with a rope and a winch, but sometimes we see one way up high on a ledge, teetering like a little china cow on a shelf. Sooner or later it will take a wrong step and fall off the cliff. More than once a cow has tumbled off the creek bank and drowned in less than four inches of water because it couldn't turn over.

One young heifer's back was broken when the neighbor's huge bull tried to mate with it. Liz hates the bull and resents the neighbor, who won't (or can't) keep it fenced, but she can't seem to keep her cows away from it. "If it would just mate with the cows, it'd be all right. But it always seems to go for the little heifers. They're too small to have those big calves."

One day we drove to the other end of the ranch to take care of a heifer who had given birth to a calf so large that her nerves had been damaged, paralyzing her hindquarters. The calf had died. The small black heifer was lying on her side with her head stretched out on the ground and her eyes closed. She looked dead, but she opened her eyes when Liz spoke to her. It was important to get the heifer lying with her head up, and if possible, to get her to stand. Cows can get pneumonia if they stay on the ground too long.

We all got on one side of her and pushed. She was warm and heavy, passive as a sleepy child. She didn't object to being rolled over onto her stomach, but she made no effort to stay there and rolled back onto her side. Bertie and I pushed her up again while Liz took hold of her hooves and bent her paralyzed back legs, folding them under her so she would stay in a semi-upright position. We took turns walking down to the river for buckets of water. She drank and drank, but only nibbled at the hay we offered by hand. We tried to get her to eat; if we left hay on the ground for her, the other cows would crowd around and eat it as soon as we left.

Liz said, "ok. Let's try to get this little gal to stand." Bertie and Liz slid a rope under her hips. They each pulled on the rope while I tried to steady her with an arm around the neck. She moaned and rocked and got her front feet under her, but when she tried to heave herself into a standing position, her back legs gave way and she lurched to the side. Bertie dodged out of the way. She was panting heavily and Liz said to let her rest before we tried again.

We tried again and then again until she finally got all four feet under her for a moment. We eased her to the ground. Liz and Bertie patted her heaving sides and praised her effort. I found myself sweet-talking her, too. She was the most pathetic creature I had ever seen.

Later that week, Liz would call me at home to report that the little black heifer had regained the use of her legs. I had never heard her sound so happy. Although she was sometimes a cantankerous companion and a rather hard boss to Bertie, my affection for Liz grew as I saw how much she cared about the cows on the ranch and the lengths she would go to save any one of them.

With hay bales, boards, and sheets of plywood, we built a windbreak around the heifer and left her while we went to tend another cow in trouble. A dun-colored cow was running back and forth on the hill swinging a grotesquely swollen udder. Either her calf had died or she had an infection. We had to catch her to find out. Liz saw a calf trailing her at a safe distance, so she must have had a painful infection, which accounted for her erratic behavior. Liz wanted to chase the cow into a corral, but there was a small opening at the end that didn't have a gate. Liz told Ambrose and me to stand at the opening to block it. Ordinarily a range cow that comes upon a human figure will automatically turn away from it. Liz assured us we were in no danger. "Just don't move or yell, or you'll freak her out even more." But ordinary rules didn't apply to this rangy cow with long, skinny horns who was out of her mind with pain.

As soon as the cow was driven into the corral, she tried to make a break for it through the only opening she could see; the fact that two people were standing there didn't even slow her down. She charged directly at us. Ambrose was standing stock still—either he was trying to do as he had been told or he was frozen with fear. Damned if I was going to let that cow run my son down. I yelled, "Oh no, you don't!" and waved my arms at her. She turned just in front of us and jumped clear over the corral fence. Liz was furious. I couldn't tell if she was mad at the cow, at us, or both. I was mad, too—at myself for putting my son in harm's way. I told Ambrose to get in the cab of the truck and stay there.

Bertie finally had the idea of catching the calf and using it to lure the cow into the truck. Once the cow and calf were in the back of the truck, we all climbed in and began the long slow drive back to Burns Ferry. The cow bellowed and kicked at the sides of the truck the entire way. She kept shifting her weight so suddenly that I thought we would tip over and plunge off the cliff. Ambrose was worried the cow might squash or fall on the baby. We were relieved finally to get the mad cow into the corral. She trotted around, tossing her head and snorting angrily.

We had to climb over a fence to get to the barn. Ambrose scrambled over easily and both Bertie and Liz practically vaulted over. I waited until the cow was on the far side of the corral and then started climbing, but she caught sight of me and charged, smashing into the fence with her horns just inches away from me. I didn't care how much pain she was in—I hated that cow.

We got the cow into the crush, a barred enclosure that clamped her in tight. We were going to have to milk that crazed animal to drain the infection. Her saucer-sized hooves slammed against the metal enclosure like sledgehammers. One of us had to hobble one leg so she couldn't kick. Bertie volunteered. I held my breath as she moved the rope slowly through the bars; she had just slipped the rope around the back leg when I felt the next kick slam against the metal. I was sure Bertie's hand was smashed against the bars, but she had timed it right. She pulled the rope tight, dragging the leg backward and tying it to the fence. Then we took turns milking her.

I had never milked a cow before. It would have been nice to learn on one of those pretty Bessies who lets you lean your head against her warm side and flicks her tail innocently in your face. I reached between the bars and tentatively grasped one of the teats. The cow bawled in pain, rolled her eyes, and jerked at the rope, trying to kick. It was obvious why she had mastitis: two of the teats were misshapen. The calf couldn't nurse properly and the udder had filled with unused milk. I squeezed a teat and managed to produce a thin stream of milk full of blood and pus. I kept squirting the foul-smelling stuff onto the ground until my arms were cramped from reaching through the bars. We took turns, and when it was my turn again, Dooley came by to see how the cow was doing. He saw me kneeling in the mud, milk and blood all over my pants, as I stripped the last teat dry. He stood and watched for a minute, gave a nod of approval, and walked off.

Ambrose remembers being afraid of Dooley when he saw him for the first time. The rancher was up on a ladder picking peaches. He was wearing baggy pants and a shirt with no elbows left, and his hair was standing on end. Ambrose said, "He'd pick a peach, hitch up his pants, pick another one and hitch up his pants, pick and hitch, pick and hitch. I thought maybe he was crazy." He eventually grew to like Dooley; later, as a pre-teen flirting with rebellion, Ambrose enjoyed being around someone who didn't give a damn what other people thought of his clothes, his language, or his behavior.

At first I found Dooley—all five feet four inches of him—scary, too. Ranchers and farmers can be intimidating. When meeting city folks, they tend to eye us with a look that says (as clearly as those I used to get from my P.E. teachers): You don't measure up. It's an attitude I've also encountered in people in the military, who seem to think that they are infinitely better prepared for life's challenges than us civilians.

I was also intimidated by Liz, but in a different way. I was amazed by the range of things she could do; not only could she lift ninety-pound hay bales and fix tractors, but she could also make exquisite beadwork and paint with oils. There seemed to be nothing she couldn't do on the ranch.

Bertie is more approachable, both because (like many musicians) she has a somewhat playful attitude toward life and because I know she makes mistakes.

One time I went with Liz to Waha to pick up a load of hay. Bertie had accidentally ordered ninety-pound (instead of fifty-pound) bales of hay and then conveniently disappeared on a tour with her bluegrass band. Dooley and a hired man (the former beatnik poet) wrestled the bales onto a hay loader that ratcheted them up to the back of the truck. Liz grabbed each bale with a hook as it came off the loader and pulled it to the top of the stack. Never in all my life have I seen anyone work so hard. Her face was flushed bright red and I thought her heart might simply explode. I couldn't begin to lift the huge bales; all I could do was push them into position once Liz got them up on the stack.

At midday the truck was finally loaded, and Liz eased herself onto her back in the grass next to the barn. I leaned back on my elbows and looked up into the sky. At almost five thousand feet, the sky was cloudless, deep, and dizzying. It seemed to be flowing toward me, dissolving into particles that floated down and disappeared out of the corner of my eye. Liz must have seen the drifting motes too, because she reached out to catch one. Her hand seemed to close on nothing, but then she turned to me and opened her fist. A tiny spider on a parachute of silk was snagged on her rough palm.

Liz drove the top-heavy truck back down to the ranch. We wound down through the trees, curving around steep ridges that fell away to nothing— pure air—just a foot or two beyond the edge of the road. Every large rock made the truck heel over for a long sickening moment—like a sailboat with its boom almost touching the water—until it righted itself. The truck groaned and backfired. I had visions of the brakes going out and us careening down the mountain until we hit a boulder and the truck became an explosion of hay and sproinging metal parts. As we bounced out of a deep rut and I bumped my head on the ceiling yet again, Liz looked over and grinned. I remembered she had invited me along because she had wanted some company on the "long, boring ride."

Ambrose comes up to me with his mouth set in a sideways twist that means he's in pain and trying not to show it. He holds out his hand, but I don't see any scrapes or scratches. Turning it toward the sunlight, I see hundreds of tiny cactus spines embedded in his palm. Bertie hands me her pocket knife, and I begin the slow process of removing them. Ambrose sits patiently looking out over the river.

Below us, green slopes dotted with yellow blossoms of cacti descend for another half a mile before smoothing out into a broad pasture along the river. On the flat, cows with calves at their sides are grazing or gathered in little coffee klatches around the salt blocks. The mulberry trees, planted here and there on the bar to attract birds, blossom white as kernels of popcorn scattered on the ground. To the south, the tops of the Seven Devils are still covered with snow. The river, shimmering in the sunlight, curves toward the west around the bulk of Ram's Head Rock, ripples into a series of rapids as it passes the bar, and then flows out of sight around a bend to the north.

I scrape the last of the spines out of my son's dirty palm. The pain already forgotten, he curls his fingers into the coarse, springy hair on top of the calf's head.

"Mom," he says thoughtfully, "Have you ever noticed the river looks like a snake?" Without waiting for an answer, he turns his attention back to the calf, gently stroking its face. The calf doesn't seem to mind his touch. "Look, he has a black spot on his nose. Do you think his mom will recognize him?"

THREE

Ram's Head Rock

It is early spring, and for some weeks now it has been cold one day and warm the next. Mud thaws and refreezes. Piles of rotting snow black with cinders line the streets of Pullman. From our house on the edge of town, the fields are a dismal striping of brown and white, bare ground alternating with old snow in the furrows and hollows of the hills. Winter wheat planted last fall is just beneath the surface, waiting for a stretch of warm weather. Soon, but not soon enough, the fields will burst into brilliant green overnight. I feel restless and out of sorts. When Liz calls asking me to help with a bonfire at the Burns Ranch, I am glad for an excuse to get out of town. Driving down Main Street, I see someone else showing signs of cabin fever: a young man in seed cap and cowboy boots tenderly maneuvers a large framed portrait of Marilyn Monroe into the cab of his mud-encrusted pickup.

Shortly after the highway crosses from Washington into Idaho, the rolling prairie comes to an abrupt end. Suddenly the ground drops away and it is a struggle to keep to the curve of the road instead of following my eye out over the valley where, far below, two rivers are shining. My eye traces the Snake River south to the Blue Mountains of Oregon and the Clearwater River east almost to Montana. As the trucks around me shift into low gear for the steep Lewiston grade, I shift mental gears for the descent into the Lewis-Clark Valley. The NPR station out of Moscow-Pullman breaks up into static and I twirl the radio knob to get a local call-in show. I listen with interest to people offering items to sell or trade.

An old woman with a firm voice says, "I need some brown eggs. Now, they have to be fresh. I don't mind if they're pullet eggs, though its pretty early for pullet eggs."

"I'm looking for some hay—mixed or alfalfa. I'm willing to trade a cord a wood."

"I still have a saw blade with a painting on it. I'll take seventy-five for it."

A caller in Orofino has a half-size violin to sell. I scribble the number on the back of my checkbook as I drive. My husband is always on the lookout for small violins for his private students.

I roll down the window as I come into Lewiston. The air is soft as rose petals but carries the rotten-egg smell of the pulp mill. Spring comes early to the Lewis-Clark Valley, which is often ten degrees warmer than the Palouse. The streets are lined with dogwood trees budding pink and white, lawns are startlingly green, and crocuses are up. I cross the Clearwater River and cross again to the Washington side of the Snake River. Driving south along the river, I keep an eye out for the bald eagles that winter in the canyon.

After traveling for miles along on the gravel road, I pull into the space next to the Burnses' garage and honk the horn several times. The sound bounces against the cliff behind the garage and echoes across the water. On this side of the river, rims of basalt are stacked in layers, bisected by a hanging canyon that spills a stream of water down the face of the bluff. The water freezes every winter into an icicle forty feet long or more. Someday I hope to be here when spring thaw comes. I've heard that the ice falls with a great booming sound, the kind of sound that shapes the space around it, so you can feel the wall at your back and the river laid out at your feet. I'm too late this year, but there is still snow on Craig Mountain, across the river.

The ranch house, hidden behind a stand of alders, sits next to Thornbush Creek, which flows out of a side canyon. I can see the outbuildings, but there is no one by the barn, and it's still too cold for Dooley to be working in the open shop. If Liz is outside she can usually hear the horn. If she is inside, my wait could be a long one. I contemplate the advantages of dynamite. Not far from here, gold miners used to set off a stick of TNT when they wanted the deaf old homesteader to ferry them across the river. That, too, must have been a satisfying sound, ricocheting off the canyon walls.

As I wait, I look closely at the grass that sticks out in blond tufts from the face of the cliff. A few blades of grass sprout between the old, each individual sprig so tiny it doesn't seem possible that, together, they will soon cover the hills in green. With the warm weather will come the fishing boats and jet boats carrying tourists, but in early spring, when the water is heart-stoppingly cold, there is almost no river traffic.

After a few minutes, I see a figure emerge from the trees. I recognize Liz by her long single braid and short, compact figure. She pulls on the oars with powerful strokes and makes quick progress across the river.

Bertie has the pickup waiting in front of the house. We drive the two miles to Ram's Head along a dirt track carved into the steep slope above the river.

In a campaign to clean up Ram's Head, Liz and Bertie have spent the last months hauling things to the hollow they use for a dump. They have carted away a couple of old cars, a wagon axle, several car batteries, a set of bedsprings, a sink, a stove, a refrigerator, and an old-fashioned wringer-type washing machine. Liz wants to have a bonfire to burn the rest of the junk—the collapsed roof and walls of the shack and whatever is buried beneath them. The job has to be done soon before the weather warms up enough to turn the road to mud. In the spring, they often can't reach Ram's Head for weeks at a time.

Ram's Head Rock is the name for a sharp bend in the river where huge boulders bunch like the massive horns of bighorn sheep. The outcropping is covered with hundreds of pictures pecked into the rock by ancient Indian tribes, ancestors of the local Nez Perce. There are bighorn sheep, elk, deer, coyotes, strange triangular humans with horns, whorls, and concentric circles full of mysterious meanings. The pictures are everywhere on the rocks, around every corner, behind every bush. They seem to be looking over your shoulder wherever you go. Several petroglyphs overlook the Burnses' old house.

It's too bad that Liz and Dooley have decided to burn it. I've enjoyed poking around the remains, looking for clues to what it was like to live here forty years ago. Headlines on 1950s newspapers behind the peeling wallpaper crow that electricity will soon be coming to Hells Canyon. Local appliance store advertisements offer deals on electric ranges and washing machines to replace wood stoves and wringer washers.

Liz is only three years older than I, and I marvel that her childhood was filled with chores more familiar to my great-grandmother than to me—cleaning kerosene lamps, hauling water, chopping kindling for the wood stove. Meanwhile, a thousand miles away, I was doing nothing more arduous than loading the dishwasher and practicing piano. We were a modern family of the 1960s. My father worked in the exciting new field of computers. Our future was bright. We neither knew nor cared that the ranching industry throughout the West was on the decline, taking families like the Burnses down with it.

It was during this time that Dooley started working a regular job to help make ends meet. Once a week, he delivered mail by boat to ranches deep in Hells Canyon. In Southern California, we felt at the very center of things as events of the late 1960s unfolded all around us. Liz saw enough on TV—civil rights demonstrations, war protests, hippies, drugs, and the race

Burns Ranch, 1951. Photo by Kyle Laughlin. 99-R-018-1. Historical Photograph Collection, University of Idaho Library, Moscow, Idaho.

to the moon—to know that she was not only miles away from everything, she was in a different decade entirely. She told me, "We didn't have the sixties in Idaho. It was just like the fifties. Guys were still greasing their hair and rolling cigarette packs in their T-shirt sleeves."

One time Liz showed me around the local history museum in Asotin, pointing out things that had belonged to her family. In a display of model tractors and windmills, there was a fish wheel Dooley had made to demonstrate how they caught salmon before the wheels were banned. The Burnses' old crank phone hung on the wall, and Liz lifted the ear piece. "I used to spend a lot of time listening in on the party line. Better than soap opera." In the basement devoted to displays about daily life, there were board games, high button shoes, and a wringer washing machine propelled by a small child-sized treadmill. I pitied the poor boy (although Liz said it may have been a goat) who had to trundle along on the treadmill while his mother fed piles of laundry through the wringer.

Liz shuddered at the sight of a rusty contraption like an oversized coffee grinder. "You don't know how many hours I spent over one of these damn things."

"What is it?" I asked.

"Cream separator. Comes apart into about fifty pieces and you have to clean and scald each one."

Intent on seeing everything in the museum, I looked through a collection of stiff, yellowed dresses, sliding each one along the wire strung across the room. Pushing aside the last garment, I was face to face with a stack of coffins left over from the building's early days as a funeral home.

When Liz wasn't doing chores for her mother, she entertained herself by watching people on the beach across the river. She learned early on that adults do a lot of stupid things when they think no one is watching. They got drunk and fought with their wives or buddies; they slapped their kids. They got their trucks stuck in the sand and ran their boats up on the rocks. They swam naked and made love on the beach, unaware of the small figure observing them through binoculars from her perch on Ram's Head Rock. Now Liz refers to all travelers on the Snake (with the exception of canyon residents) as "idiot tourists," whether they are locals who drive out from town in a pickup with a fishing rod and some beer or out-of-staters who come down the river on a guided tour of Hells Canyon. Some of them shot at eagles, smashed the heads of rattlesnakes with rocks and cut off the rattles, and spray-painted their names over the petroglyphs. Sometimes locals beach their boats to eat lunch on the rocks or to use the outhouse, but if no one is around they may throw empty beer cans around, shoot something up, or tear something down.

Last summer Liz had encountered a woman walking along the beach carrying a huge armload of poison ivy. Asked what in hell she thought she was doing, the woman said she was setting up camp and was going to use the leaves to cushion her sleeping bag. Liz went up to the house for a bar of soap and told the woman to go wash in the river and in the future to try not to be so incredibly stupid. Actually, she didn't say that, but I knew the voice she used when she thought you had done something really dumb. It is the same tone of voice she uses when I let the fire go out, or put the wrong bridle on the horse, or ask too many questions.

My fascination with Liz's old house is shared by some of the "idiot tourists" who stop at Ram's Head to look at the rock art. Liz is afraid that sooner or later someone is going to get hurt climbing around on the rotten floorboards. She also hopes that without the ruins of the shack, the old homesite will look a little more presentable. Liz knows that one reason Ram's Head has been so heavily vandalized is that it looks abandoned.

Over the years, the old Ford in which Liz learned to drive when she was eleven years old has become so riddled with bullet holes that the fenders look like cheese graters. The cabin that Dooley made into an art studio for her mother has been broken into so many times that Liz said, "I might as well put in a revolving door." The vandals always come in through the south window, leaving muddy bootprints on the windowsill. I wonder if they are disappointed by what they find. There isn't much to see—an old cot, a wood stove, a few cans of food with peeling labels, and the moldering remains of a pack rat in a sprung trap.

From across the river the barn, with its roof glinting in the sun, looks deceptively new. It is only up close that you can smell the dry-damp smell of decaying wood and hay. The barn, built in the 1890s, was close to collapse when Dooley rigged up an array of winches and pulleys to pull the two sagging sides back together. The walls were reinforced with every solid board they could find on the ranch. Then Liz and Bertie climbed up on the roof and hammered sheets of corrugated metal over the gaping holes.

While Bertie tends the fire, Liz and I pull up the tar-paper roof from the collapsed shed near the house. It had been used to store tools and spare parts and the usual overflow of a household—outgrown clothes, well-thumbed magazines, old kitchenware and canning jars. There are sharp pieces of broken metal, shattered glass and crockery. I'm afraid something sharp will go right through my soft shoes, so I use the boards as a walkway.

Liz crunches into the midst of the scree in her thick-soled boots and hefts a couple of unbroken quart jars. "What do you suppose is in here?" A clear liquid sloshes back and forth, leaving a brown scummy line against the inside of the glass. She grins. "Some of great-grandpa's moonshine, maybe? Gasoline? Well, I'm not going to find out by throwing them on the fire." She sets them aside to take to the dump.

"Looky what I found," she cries, pulling a finely made wooden level from under a collapsed shelf. Turning it over she shows me the name carved on the back. "Hiram Burns. My grandpa made this." She holds it up to admire the movement of its bubbling eye. "It's not even broken." Liz wraps the level tenderly in a rag and puts it behind the front seat of the pickup.

I have on a stocking cap and long underwear beneath my clothes, but I am cold whenever I stop moving. I look with envy at Liz and Bertie's insulated coveralls and caps with fleece ear flaps. Still, it feels good to be outside doing something vigorous and useful. After dumping the wheelbarrow, I stand next to the fire to warm up. The smoke stinks of burning tar and makes my eyes water. Turning away from the smoke, I feel the wind blowing out of the further reaches of Hells Canyon, still with the bite of winter. To the south, before the river makes another turn and the canyon walls cut off any further view, there is one bright glimpse of the high mountains covered with snow. As I take the wheelbarrow back for another load, I try to picture Ram's Head in just a few months—the house finally gone, the ashes hidden beneath new grass, and the pasture covered with flowers.

Sifting through the debris from the house and shed, I keep finding things I think Liz might want to save. "Keep this?" I ask. After a quick glance or a passing "Oh yeah, I remember that," she says, "Burn it." Liz doesn't want any of the odds and ends of her childhood that we turn up; she is more interested in finding salvageable boards to use for her endless fence mending. I marvel at how easily she parts with the kind of things I keep in my garage in carefully packed boxes labeled "childhood memorabilia." Now that the beach community where I grew up has been changed almost beyond recognition, every scrap of childhood still extant is precious to me. Maybe its easier to destroy your own past than to have others do it for you.

We throw into the fire a bingo chip, aprons, spoons, teacups, saucers, a 1964 postcard from Glacier National Park, a flatiron, a parents' manual for teaching a child to swim, toothpaste, slippers, a comb, high-heeled shoes, a child's moccasin, wooden alphabet blocks, a single playing card.

We burn 1961 income tax instructions, feather pillows, canceled checks, coloring books, a jar of Vicks Vapo-rub, hardback copies of *The Arabian Nights* and *Little Black Sambo*, rules and regulations for riverboats, and a dress for a paper doll. I show Liz a 1957 Sputnik button and then slip it into my pocket.

It feels entirely appropriate to be working with fire here. Fire has always been a part of life at Ram's Head. Liz's earliest memory is of watching their home burn to the ground. The two-story house, which sat in the middle of what is now the pasture, was built in the 1920s when the land belonged to Liz's uncle. There is a photo of the house taken not long after it was built. Neat as a midwestern farmhouse, it looks out of place stuck square in the middle of Ram's Head Bar without a scrap of shade around it. All that's left now is the cellar—a hole in the ground with stone steps leading down into the dark. In the summer I give the cellar a wide berth; rattlesnakes often come out of the hole to sun themselves on the rocks.

Liz's earliest memory is harsh and singular, while mine is a benign composite of many summer days in Southern California: walking to the beach on a fine blue and white day with a breeze just soft enough to lift my hair; I am carrying a red metal bucket and looking forward to seeing the colorful beach umbrellas against the white beach.

After the old house burned, Dooley cobbled together two sheepherder shacks brought down on skids from Craig Mountain. One shack was used for the kitchen, the other for a bedroom, joined by a small living room built of scrap lumber. Later they brought in a camping trailer, put a lean-to roof over it, and used it for a second bedroom. Dooley put the house against the back of Ram's Head Rock so the rock blocked the wind in the winter and provided shade when the sun was in the west.

The hay field caught fire several times during Liz's childhood. She remembers the fires as especially terrifying because her parents got her out of harm's way by setting her in a boat on the river. "Can't watch a kid and fight a fire at the same time," Dooley said. He would tether the rowboat to a rock on shore and push the boat out into the current. Four- or five-year-old Liz would sit very still in the boat as she watched the dark silhouettes of her parents beating out the flames with wet feed sacks.

Once Liz showed me her photo album—a hodgepodge of black-and-white photos with serrated edges and old Polaroids with bad color like a television picture in need of adjustment. There was a blurry snapshot of her grandmother showing off a fine head of cabbage. "That's me," she said, pointing at the cabbage—six-month-old Liz almost hidden by a ruffled dress billowing up around her. Grandma Louetta died when Liz was thirteen, but she remembers many of her sayings. When her grandmother dismissed someone without gumption by saying "they got no feet," Liz thought she was talking about one of the rock drawings of a human figure without feet. Being able to stand on your own two feet and take what life dishes out to you was important to Louetta and the other old-timers in the canyon. In Idaho, you have to be able to "stand up to the rock."

"And here's Pop in his Gary Cooper days," Liz said, pointing at a picture of her father squatting on his boot heels in the dirt, baby Liz propped on his knee. With his hair slicked back and his western shirt with snap buttons white against his dark skin, he looks cocky and handsome. Another photo shows Liz as a toddler cooling off in a plastic pool, her mother standing nearby, smiling. Gina stands with her hands on her hips and her chin up— a stance that seemed to say to life, "Give me all you've got." I could see, finally, what other people had told me about Gina. To a person, they said, "She was really something."

In those young photos, Liz looks adored and trusting. But when she was three years old, one of the ranch dogs bit her savagely on the face. She doesn't remember the accident, but Dooley remembers that day all too vividly. "It was that goddamned dog. Just about a five-year-old dog. Had it around as a pet. We was moving cattle. Gina got out to open the gate and she left Lizzie standing by the car. I got out and was runnin' the cows down through there. What made that dog bite that kid, I don't know. It run up and grabbed her by the face. One tooth went into that eye and the other went in and broke some of her front teeth loose. It was a *terrible* mess. Of course, we doctored and doctored and *doctored* her. The eye healed up, but it wouldn't line up with the other one. Over time, why, she finally trained them enough to where they is both looking in the same direction, but it was a *terrible* problem for her. Still is."

The scars took years to fade. There were no pictures taken of Liz for several years. When she reappears in family photos at the age of eight, she looks uncomfortable and self-conscious, as if she has been coaxed in front of the camera. She admits, "I was always hiding my face." In the pictures, her blind eye is turned inward, as if she is looking at something deep inside herself.

When Liz started school in Asotin she was so shy and unlearned in book matters that the teachers thought she was retarded. They did not consider it remarkable that she could tell dozens of cows apart by their markings, that she knew which plants were good to eat and which could kill livestock, that she could catch fish with a string made of grass and a piece of bent wire. The enclosed classroom was stifling to her. The kids didn't like the way she looked at them—or didn't look at them (they couldn't tell which)—with her strange eye, and they often beat her up. On particularly bad days, she would just run off and spend the rest of the school day down by the river.

Once Liz showed me how she made string. She pulled up a piece of strong grass called "Indian hemp" and split it into two strands. Then she wound one around the other, twisting them in opposite directions so that when the string tried to unwind, the two strands would resist each other. She handed me a piece of grass and I managed to get the string going for a few inches, and then, like a piece of unraveling yarn, it came apart in my hands. Liz quickly wound a string as long as her arm and flexed it to show how strong it was. "See, all you need is one of these and a bent piece of wire and you're in business."

It seemed incredible to me that someone could catch a fish with such a setup. I had been fishing perhaps a dozen times in my life but never caught anything, except for one time on the Gunnison River in Colorado when I snagged a sucker by the eyeball. I pulled out the hook, taking the eyeball with it, and threw the fish back in the river. I haven't been fishing since.

When Liz was a teenager, the roof on one of the sheepherder shacks collapsed. Rather than rebuild the structure, the Burnses decided to move to the other end of the ranch, where there was a house that had been abandoned for several years. They just walked away from Ram's Head, leaving behind anything that wasn't worth the trouble of hauling to Thornbush Creek.

It was difficult to imagine living in a house so insubstantial that it could simply fall down one day. The home I grew up in—a large Spanish Colonial house made of stucco and brick—seemed as solid as a castle. It even had a wrought iron gate across the drive and a stained glass window of a castle on the stair landing. The house sat at the intersection of Broadway and Knob Hill in Redondo Beach, a block and a half from the ocean.

My parents bought the house from three old maids whose father had built the place in the 1920s. They hadn't changed a thing in over four decades. The wallpaper was dark flocked green, the wool carpet a swirling floral pattern. The light switches were weird double-button affairs: the lower one popped out when you pushed the top one in. At the back of the house on the second floor was a long room with twenty-one windows that opened like little doors, offering an unobstructed view of the eucalyptus trees and red tile roofs of the Palos Verdes Peninsula in the distance.

From above, the pedestal bird bath in the yard made a concave circle like a Smarties candy whose indentation perfectly fit the tip of my little finger. Just visible between the garage and the pepper tree was the red peaked roof of the two-story playhouse my father built. I spent many hours looking out on that view while I was supposed to be ironing pillowcases. It was a boring, pointless job, but when I was finished, I had a neat, clean-smelling pile of pillowcases, which I carried to the landing and put away in a linen closet full of enough sheets for a small hotel, stacked on shelves you could climb like stairs to the attic door just above.

The backyard was large for the beach cities, though it seems small and plain now compared to our overgrown yard in Pullman. One summer my six siblings and I put on a circus in the yard and sold tickets to the kids in the neighborhood. I sat in the playhouse dressed like a gypsy and waved my hands over my crystal ball—a crystal doorknob from the laundry room door, gone purple with age. I bent over my friends' palms, tracing their long life lines, the little offshoots that meant they would have four—no, five—children.

When I was away at college my parents sold the house. All seven children were gone, and my mother was tired of rattling around in the empty house like a button in a Band-Aid box. They sold it to an architect who gutted the kitchen, tore out the antique sideboard in the breakfast room, knocked down walls, and (most disconcerting of all) reversed the staircase. The architect and his wife have no children. They live quietly in the thirteen-room house that used to be overflowing with kids and commotion. I imagine evenings with just the two of them—the wife reading a magazine, the room silent except for the sound of turning pages and the hiss of the gas jets in the fireplace, the architect in his leather recliner musing over what to dismantle next.

The house on Knob Hill was just a few miles from where my great-grandfather Johnston, a wealthy newspaper publisher, built a grand house with timber from a shipwreck on the rocks at the end of the Palos Verdes Peninsula. My grandma Smokey spent her teenage years in the house—a spoiled only child waited on by servants. My father also grew up there. As a small boy, he used to swing from the huge open beams on the porch. In the evening, his grandmother would put aside her sewing to remove the slivers from his palm, pricking gently with her sewing needle at his hand, held still and flat like he was feeding a horse.

On a visit to Southern California several years ago, I suggested my dad take me to see the house, which had passed out of our family in the 1950s. He was distraught to discover that the house was gone. He stood there sadly, looking at the cantilevered thing with tinted windows that has replaced the graceful house. "Dad, let's go," I said. But he caught sight of a garden wall still standing at the edge of the property.

"My father built this," he said, touching the warm brick. "Just after they were married. Eskil's hands were so huge, he could pick up four bricks with one hand." His own hand resting on the wall has the same broad palm that my brothers and now my sons have inherited as well—the hands of a Swedish bricklayer.

My father died not long after he took me to see the house in Hermosa Beach. I regretted ever asking him to take me there; I felt that in my eagerness to claim my own history I had taken away part of his. I wish he had died believing his grandfather's house still stood high on the hill, the ship's lantern on the porch a beacon guiding a child home.

On our daily walk to Lutheran school, we used to cut through empty lots covered with sand and iceplants that gave a satisfying juicy crunch underfoot, like walking on celery. But the real estate boom of the 1970s and 1980s made the lots too valuable to stay empty. They've been filled in with condos, espresso stands, and specialty shops. Single-story ranch houses built in the 1960s have been replaced by two- and three-story houses that vie with each other for a view of the beach. Run-down bungalows that once housed poor Mexican families have been retooled into micro-mansions with wide-screen TVs and elaborate barbecues.

The elementary school has been turned into a community center and city museum. (I imagine it filled with crayoned maps of Bolivia, old SRA reading booklets, deflated foursquare balls, my science project on the Möbius strip.) My first love—a bad boy with an insouciant blond flop of hair over his eyes—cut our names deep into the bark of an old maple tree behind the kickball diamond. For years, I would climb the tree and trace the crooked heart with my finger, feeling how the jagged letters had worn smooth with time. The maple tree has been there as long as anyone could remember, and I thought our names would be linked forever, too (a mortifying thought once I stopped liking the boy). Now the tree is gone and the playground is a parking lot.

While the explosion of building was going on, families could not afford to live in Redondo. The beaches, once crowded with children popping seaweed bulbs, digging for sand crabs, and chasing sandpipers were empty except for joggers and the occasional woman sunbathing in hat and sunglasses with a book on her lap. Now the young professionals who remodeled the houses and resodded the lawns are having kids.

I wonder if in a few years these children will be like us, spending their childhood beneath the watchful yet disinterested eye of the lifeguard, staying in the water until their fingertips turn blue, then lying on their stomachs on the beach, slowly paddling their arms and legs like turtles in the warm sand. Maybe teenagers will run on the sand as they used to, the boys snapping at the girls' legs with the wet twisted corner of a towel, the girls screaming in mock anger and real pain and showing off the red welts on their thighs. But somehow I doubt it. These children lack what we, in the 1960s, had plenty of—time. We squandered massive amounts of time at the beach: entire summers; fall afternoons when the hot Santa Ana winds drove us to the water after school; foggy winter days perfect for beachcombing; and later, as adolescents, nights on the cool sand with a boy with the smell of Doritos and beer on his breath.

The beach worked its way through us, grain by grain, a steady stream of all those hours on the sand until the beach was part of us. It takes that long: whole beaches of sand, layers and layers of rock. The Snake River is a popular route for rafting, and whitewater adventure companies whisk hundreds of people down the river each summer. But what can anyone absorb on one pass through a canyon as old as this? With each visit here,

I lay down a sheet of memory, thin as onionskin. They add up over time, but I still have barely begun to know this place.

Here on the ranch, the past is allowed to fall away incrementally. The shack caves in and mice make nests in piles of old clothes, the rock wall at the base of the bluff tumbles down stone by stone. Up on the mountain, moss grows on an old cabin roof. The walls lean and lean before they finally give. But in California the past has disappeared in a rush, burying my childhood in a Pompeiian flood of concrete.

Darkness falls quickly in the canyon as soon as the sun drops behind the western rim. We stand around the dying fire, kicking at the few things that still smolder, sending up sparks that glow briefly in the dusk. Two owls with big, blocky heads fly out of a crack in the canyon wall and on across the river. I think of the night, several weeks ago, when I had been on the way to a writer's workshop at Wallowa Lake in northeastern Oregon.

I drove through tiny Anatone with its sign at the edge of town: Population: 36 people, 11 dogs, 10 horses, and 15 cats. I started down a lonely stretch of road lined with dense forest. Out of the darkness, a barn owl appeared and hovered over the car. Its wings fully spanned the width of the hood, and their feathered undersides were so pale they seemed to emit a light of their own. I remembered Liz saying once that an encounter with an owl is a sign of an impending spiritual change. I don't particularly feel the need for a spiritual conversion. I would be satisfied with being able to tie a knot that won't slip, with knowing the name of the bird with the blue head and orange breast that sings on top of the mulberry bush, with being at the ranch enough for this place to seep so deep into my bones that it will stay with me no matter how far away I may go.

FOUR

A Childhood Out of Time

Every child feels herself at the center of the universe; we think the moon follows us alone. But for my six siblings and me growing up in Southern California in the bright and booming 1960s, we knew it to be true. We saw our lives reflected in movies and TV, on the radio, in commercials and on billboards. On television, everywhere—even Mars or the Soviet Union— looked like Southern California. The entire planet was just like this: sunny, cheerful, and safe; and if it wasn't, the adults seemed to say, well then, it ought to be.

It seemed only fair that TV was our mirror image, given that we never saw ourselves in books. All the children we read about lived back East. In winter they wore hats with pompoms on the strings and went sledding; in summer they went to camps with idiotic names. The West appeared in children's books only as the exotic setting for a summertime adventure: catching cattle rustlers while vacationing on a dude ranch or discovering dinosaur remains while visiting Uncle Bob.

It was the image of leafiness that separated East from West. The children in books were always raking leaves, playing beneath big shade trees, or riding bikes down lanes covered with green canopies. Our trees were dry and spiky. They grew taller with no apparent effort, like our dolls that grew hair when we pulled with a steady pressure on the middle of their scalps. Back East, the trees fit the scale of the houses, as if they had grown up together. Here, it seemed the smaller the bungalow, the taller the palm tree. We had few deciduous trees or migrating songbirds to mark the seasons, no snow, no real spring. It could be sunny and warm at Christmas and foggy on Fourth of July or vice versa. There seemed to be no pattern to it, no rhyme or reason. With little sense of time passing, we lived in the present tense. Now was a good place to be.

Our magic kingdom of Southern California was under the benevolent

The Johnston family at Venice Beach, 1908.

reign of Uncle Walt. He spoke to us every Sunday evening in a slow, kindly voice that we trusted the way adults trusted Walter Cronkite's voice. He was not the only one who spoke to us baby boomers. Owing to sheer numbers and the spotlight put on the younger generation in the wake of Sputnik, children were treated as if they counted. We were told it was up to us to continue the space race someday, just as we imagined we would carry on the endless war in Vietnam. We even had an assigned part in the nation's response to nuclear attack: we would all be hiding under our desks, heads tucked between our elbows.

L.A.'s aerospace industry kept the country safe at the same time it was flying lucky people all over the friendly skies. Many of our neighbors worked at Lockheed, Martin Marietta, TRW, and the other huge facilities that lay a few miles to the east. My father was an electronics engineer for a small company that did work under government contract. Electronics engineering was evolving into computer engineering, and my father was proud to be part of an exciting new field. He explained computers to us by saying they were good at counting things really, really fast. It seemed like a silly thing for a whole roomful of machinery to do, but he assured us that computers would be capable of all sorts of amazing things by the time we grew up. We imagined ourselves consulting mini-computers on our wrists as we flew about with airjet packs on our backs.

The 1960s were our family's golden age. The younger kids were old enough to join in family activities and the older ones hadn't yet left home. We were more in sync with the times than any time before or since. Our living room was full of up-to-date furniture, including a butterfly chair and the kind of pole lamp rarely seen nowadays outside of a dentist's office. My father often came home with a brand-new record for us to listen to during dinner. We memorized Bill Cosby's "Chicken Heart" routine and studied the lady covered in whipped cream on Herb Alpert and the Tijuana Brass's new album cover. Our mother had not yet discovered women's lib or health food and she still baked cookies for us after school. I was in no hurry to grow up; if my brothers stayed children, they would never have to go to Vietnam.

We lived on a wave that never broke, the future curling just ahead, the past receding, flat and indistinguishable, behind us. For children to have little vision of the past or future is normal, but we were surrounded by adults who shared the same view. People came to Southern California to escape. The last thing they wanted to talk about was where they came from and why, and who they had been when they left home. The father of my best friend, Moana, lived through Pearl Harbor; after the war, he settled—I imagine with a sigh of relief—into a job at the unemployment agency, where he worked until his death thirty years later. A friend who looked like a typical Californian (bleached blond hair, nose job, great tan) was born in Los Angeles just a few months after her parents fled the 1956 revolution in Hungary. For these parents and for our neighbors—the elderly Austrian couple down the street, the Greek woman who did alterations for my mother, the Mexican families who lived in the run-down neighborhood by the pier—today was better than yesterday and tomorrow was going to be even better. What was there to talk about? Go outside and get some sun.

We returned to L.A. in 1964 after four years in Washington DC; the place seemed both new and strangely familiar. We lived in Redondo Beach, one of three cities that hug the south end of Santa Monica Bay. Redondo had been a fashionable resort town around the turn of the century when it was the last stop on the popular Balloon Line. L.A.'s unparalleled electric rail system brought vacationers to the beach to stay at the two-hundred-room Hotel Redondo overlooking the pier.

Surfing was part of the city's identity from the very start: this was where surfing was first introduced to the United States. Beginning in 1907, the famous Irish-Hawaiian surfer George Freeth put on twice-daily exhibitions

Redondo Beach Pier and Lightning Racer Roller Coaster, 1919. Title Insurance and Trust Company (Los Angeles) Collection, Redondo Beach Historical Commission, Redondo Beach, California.

in front of the Hotel Redondo, barreling down the waves on an eight-foot-long wooden board.

By the 1960s, although the Beach Boys were still singing about Redondo's surf, there was little left of the city's former grandeur. The much-loved Big Red Cars of the trolley system had disappeared as freeways cut across their routes and siphoned off their passengers. The Hotel Redondo had been torn down after only thirty years. The Lightning Racer roller coaster was demolished next. The Plunge (billed as "the largest heated saltwater pool in the world") was razed in the 1950s. The city lapsed into somnolence, becoming a quiet family town that wasn't as stylish or hip as the other beach cities. It was, however, more peaceful than crowded Hermosa or Manhattan Beach, with large turn-of-the-century houses on the cliffs above the beach and wide streets lined with pencil-thin palm trees.

Our parents had gone to the local high school, but we had no sense of a long-standing family connection to the area. We felt as loosely planted here as the other families who were swept in with the surging aerospace industry. We were unaware that all four sets of our great-grandparents had arrived in California between 1880 and 1906. In the West, a hundred years is a long time. In California, one hundred years is beyond imagining. Who could even conceive of a Los Angeles with a population of fifty thousand people, as it had been when the Inglis family arrived in the 1880s? There were that many people in Redondo Beach alone by the mid-1960s.

Having a long family history in Southern California was like being a root vegetable among a bunch of hydroponic tomatoes. Our history as a family in the Southland was irrelevant and rarely mentioned. The overwhelming majority of people in L.A. had arrived after World War II, and more people arrived every day. Unlike many other parts of the country, where once an outsider, always an outsider, here it didn't matter if you arrived a century ago or the day before yesterday. If you were white and spoke English, you were accepted as a Californian, no questions asked.

My mother's parents could have provided us with a sense of family history, but our access to them was limited. The Rhodeses' interest in grandchildren was simply too thin to be spread among seven children. Their natural inclination would have been to pick one or two of us as favorites. My mother tried to thwart this by allowing them to see us only en masse. On their infrequent visits, my mother would check us over to make sure we were wearing clean clothes and then usher us into the living room. One by one we stepped up to be interviewed by our grandfather. How are your grades? What are you doing to help your mother around the house? Are you saving part of your allowance? Praising us or chiding us according to our answer, he would select a quarter from his coin purse and place it in our palm, handling even this small transaction with the gravity of a true Depression-era man. We were then shooed out of the room.

The Rhodeses lived in a section of town called the Riviera. When we were old enough to ride our bikes there, my sisters and I took turns cleaning their house. It was the only time we were allowed to visit our grandparents' house unsupervised, and we had to work in pairs. My older sister took on the heavier jobs such as vacuuming and mopping. I wandered around with a feather duster, supposedly dusting my grandmother's porcelain figurines but mostly straightening and restraightening the lavish coffee table books so I could sneak a look at them. I was able to glean much of their contents by looking at the pictures and reading the captions. From those books, I learned things about California history that weren't taught in school. The past clicked by in separate episodes like scenes in a Viewmaster: first the Indians, then the Spanish, then the Mexicans, then the land speculators and developers, and now Hollywood.

I read gruesome accounts of how the Indians had been mistreated by the

Spanish at the missions. I read the first L.A. laws passed under American rule. Never having seen a dark night in Los Angeles (the night sky was usually a kind of boiled orange color), I was interested to read that town residents were required to put a light in their door for the first two hours of every "dark night." Another law authorized the formation of chain gangs from the city jail (mostly Indians serving time for public drunkenness), which could be auctioned off as day laborers. Why, they were almost like slaves!

There were photos of tiny men standing beside huge hoses aimed at the exposed hillsides of the Mother Lode country. I leaned close in the dim room to look at their faces, trying to see what kind of people would wash away a whole mountain for a handful of gold. It was intriguing to discover that L.A.'s water had been stolen from the farmers in Owens Valley. But the machinations of the city leaders were impossible to follow, and I was stymied by a basic incomprehension: How does one steal water?

The scandal of the silent film star Fatty Arbuckle was the most puzzling of all. I returned over and over to the story of the three-hundred-pound comedian charged with a young actress's death after a mysterious bedroom encounter. (I finally concluded he squooshed her to death.) California history seemed to be made up of shameful and sordid incidents. Something you weren't to be caught reading when an adult walked into the room, especially when you were supposed to be dusting. No wonder nobody talked about it.

My grandparents' house was so still, their life so ordered, that I could easily believe my mother's claim that her childhood in Manhattan Beach had been so boring that she didn't remember it. The only childhood photo she had kept showed her at age twelve with a cast on her foot, grinning. She had broken her ankle running on the beach: at last, something eventful had happened in her life.

"What about World War II? Wasn't that exciting?" I asked. She mentioned victory gardens, blackouts, scrap metal drives. I wanted more, but my mother was not a storyteller. Her communication tended to the nonverbal: a sigh, a raised eyebrow, even the clack of her sandals against the wood floor signaled her mood. But she managed to unearth both a half-used book of ration stamps and a stack of children's magazines called *Story Parade*. I read the magazines slowly, as if I were a child on the home front waiting impatiently for the next installment of "Champions Don't Cry" or "Bill and His Bulldozer Help the Fighting Seabees."

My mother had a more precise memory of her teenage years, but I had little interest in the lovely gowns the rich girls wore to the high school dances (envy of the horsey set in Palos Verdes was an ongoing theme in my mother's life). Her time was filled with school and music; she practiced several hours a day, playing piano, organ, oboe, and—an image that never fails to fill me with delight—marimba. When she married my father at age eighteen, she had never been outside Southern California.

Perhaps this was why my mother took to the road so readily as an adult, despite having to pack seven children and enough food for two to three weeks at a time. Among our friends and neighbors, our family was an oddity for its wide-ranging travels. Nobody else in Redondo seemed at all interested in getting out of California, although occasionally someone we knew would drive down to Baja to camp on the beach, or fly over to Hawaii to lie out on . . . another beach. There was no need to leave the state, they argued, when we have Disneyland, Hearst Castle, Yosemite, Lake Tahoe, the Sequoias, and San Francisco. We've got the highest mountain in the whole lower forty-eight. We've got the hottest, lowest desert of them all. We've got more beaches than you can shake a stick at. Why in the world would you want to go anywhere else?

On our camping trips, as on other family outings, we were dressed in matching outfits. In cold weather, red hooded sweatshirts; in warm weather, sunsuits that tied at the shoulder. For going to the ballet at the Dorothy Chandler Pavilion, my mother put us in matching velveteen and taffeta dresses, my crewcut brothers in blue suits and clip-on ties. On one cross-country trip, all five girls (six including my mother) wore pink candy-striped sundresses she had sewn from a Simplicity pattern.

We drew comments wherever we went, especially when my father had us line up for a photo on a scenic overlook or some other conspicuous spot. Our parents were enormously proud of us and loved to show us off. We were stair-step children, all close in age, brown-haired except for our elfin youngest sister, Ingrid, blonde as her Swedish namesake. Many people, even those who had known us for years, found us indistinguishable, which didn't trouble us a bit. As teenagers, we struggled to define ourselves as individuals, but as children, our group identity was a source of security.

The size of our family often prompted strangers to ask, "Are you Catholic? Mormon?"

This puzzling question left me tongue-tied. What's religion got to do with it? I wondered. I would blurt out, "No, we aren't anything."

We were no particular religion or ethnicity. If my parents had faith in anything, it was education, and we were baptized only so we could attend Lutheran schools that held to a higher academic standard than the public schools. No traditions had been handed down to us from previous generations, and few stories. We had infrequent contact with our extended family, and my parents seemed satisfied with this arrangement. They wanted to create a family from scratch, with its own traditions and rules, unblemished by any contact with the past. We were nothing but the Freeman family, and so they made that everything.

My parents' vision of the ideal family was an imperfect one, given that neither of them had grown up in a close, loving family. My father's standard reply to any question about his parents was, "I don't know. I hardly knew them." My father would talk about his childhood only if pressed, and even then slowly and painfully. We knew he had spent seven years at a boarding school in Los Angeles, about twenty miles from his home in Hermosa Beach. "But why did they send you away to boarding school?" we asked at intervals throughout our childhood, trying to picture what it was like to grow up without parents, siblings, or a home. "Where was your mom?"

"She was in Sacramento." The explanation that followed sounded rehearsed, as if it were something he had heard repeated many times. "It was the Depression. That was the only place she could find a job."

We knew that Sacramento was where the governor lived. It was far away. "Where was your dad?"

"He went back to San Francisco after the divorce."

Divorce was unfamiliar to us, and it amazed us that someone could simply remove himself from a family like checking out of a hotel.

My father was only two when his parents separated. He moved from their small house into his grandparents' much larger house next door, where he was cared for by a maid. The Johnstons obviously loved this little boy with a confident grin and a penchant for taking things apart; in forty years' worth of photos, the only ones in which they are smiling are those taken with their grandson. My father had fond memories of them both. But his beloved "Babu" died and his grandfather developed heart problems. When it came time for him to start kindergarten, the only solution appeared to be to put young Peter in boarding school.

We found this so incredible that we sometimes asked again, "*Kindergarten?* Five years old?"

My father would always think for a moment, as if he might remember it differently this time: maybe he *had* been older, maybe his mother *had* visited him every single weekend and he just forgot. "Yes, I must have been five, because my first memory of my mother is when she came to visit me on my sixth birthday, and by that time I'd been at school for a while."

For my sisters and me, boarding school had all sorts of dramatic possibilities. We longed for our father to tell us stories about nighttime escapades with a daring and loyal best friend with a name like Kit. But it hadn't been like that, he said. It was all right at first, when the California Children's College was in a lovely old house in West Los Angeles. But later it was moved to a barracks-like building previously occupied by a military school. His grandfather's visits grew more infrequent as his heart condition worsened; then he stopped coming entirely. His mother rarely came to see him, even after she moved back from Sacramento. His father did not visit him at all.

The most difficult thing for my father, he said, was that most of the other children were allowed to go home on weekends and holidays. "Sometimes I was the only child there over the holidays." Throughout my childhood, I was haunted by the image of a small boy in school uniform, wandering the halls in a fog of loneliness. Each summer, he and a handful of other similarly orphaned students were bussed to the school's summer camp in the San Gabriel Mountains. There my father learned the consoling power of nature, and for the rest of his life the mountains were where he felt most at peace.

My father never said a word against either of his parents and their abandonment of him, yet his silence was condemnation enough. His family history was a closed book. Even to express interest in either of his parents felt like a betrayal. Nevertheless, I was fascinated by his parents and regretted that neither of them were around to answer my questions. His father lived in Mill Valley, across the bay from San Francisco, and visited us only two or three times that I can recall. He was big and handsome in a typically Scandinavian way, and he had a courtly charm that must have caught my grandmother's eye the first time she saw him, at a town meeting on St. Patrick's Day 1925.

On one of his rare visits, Eskil took my family out to dinner at a Swedish smorgasbord. Scooting behind the chairs, my little brother ran into the lit end of my grandfather's cigar, burning his eyeball. Eskil's entire visit was spent sitting silently with us in the car while our parents were in the emergency room with my brother. I sat next to Eskil holding his giant hand. There were things I wanted to ask him, but he had a look on his face of such deep sadness that I dared not.

My father's mother had moved to Alaska Territory in the 1950s, and her letters to us were chock-full of northern lore. By the time I was in sixth grade, Alaska had long since entered the union, and I knew more about the forty-ninth state than I did about California. Even if our L.A. relatives had been inclined to tell us stories, perhaps we wouldn't have listened. As children we were naturally more interested in the story of a pack of wolves chasing our grandpa Scotty than in how our great-grandmother Taylor got the laundress to stop boiling the wool underdrawers.

When Wendella Valeria Johnston Freeman remarried and moved to Alaska, well before I was born, she left her old life behind. (It was her fourth marriage—the first ended in an annulment, the second in divorce, and the third had left her a widow.) At age fifty-seven, it was a new start for her, and she embraced it wholeheartedly. She even took on a new name: Smokey Johnstone. (Her oil paintings sold better, she claimed, if the signature did not identify her as a woman.) Smokey was a nickname she had had as a young woman when she had been known for her hot temper. And she was happy to change her last name yet again: it amused her that Scotty's last name was just one letter different from her maiden name of Johnston. Smokey fell so in love with her adopted state that she left it only once in the three decades before her death.

Smokey left all of her family photos, furniture, and mementos with us. My father kept track of it through several moves, inventorying the whole collection to reassure his mother that it was, indeed, all still accounted for. Cloisonné-on-brass vase. Tea set in a padded wicker case. Steamer trunk with the initials W. V. J. stenciled on the side. Each item was accompanied by a neatly handwritten note explaining its significance. The press badge was from the First International Press Association Conference in Hilo, Hawaii, in 1920, where she had been one of only three women delegates. The silk shawl had been brought back from China by one of her Yankee trader forefathers. The brass candelabrum was used in her wedding at the grand Mission Inn in Riverside, where Teddy Roosevelt had once stayed.

There were a few items from my father's childhood: school papers with the strange name of Peter Hoberg, my father's name until the age of twelve, when he was adopted by his new stepfather; and the flag that was draped over the stepfather's casket four brief years after he came into Peter's life. I found it remarkable that my father kept these painful reminders of a past he would rather forget; a less generous man would have sent it all to the dump. But he thought his mother might be right: one day these things may mean something to his children.

As a teenager, I was drawn to the garage full of wooden crates and furniture covered with a sticky glaze of dust and damp salt air. I spent many hours refurbishing Smokey's steamer trunk, overcoming my aversion to power tools long enough to knock the rust off the hinges and clasps. It became the centerpiece of my bedroom, an altar to the old and the wild. At various times, it held a rusting Colorado license plate, a packet of columbine seeds, my father's silver christening cup, an abalone shell so old the iridescent layer was flaking off, Smokey's cloisonné vase, buttons made from elk horn, and an ivory-handled pen knife.

I slept beneath a quilt made by my great-aunt Mary (a pathetic threadbare thing, but the only quilt we had in a family not known for its women's handiwork) in an ornate four-poster bed my wealthy great-aunt Marjorie later reclaimed ("for her gigolo," my mother fumed). I brushed my hair before my great-grandmother Grace's oval mirror and read by the light of her brass floor lamp. My parents seemed amused by my enthusiasm for antiques when their other children's bedroom decor tended toward beanbag chairs and lava lamps.

To me, the antiques were reminders that there had been a time when the South Bay was still a real place. As young people, my grandparents had caught the grunion that came ashore at night to spawn, pried abalone off the rocks with special curved knifes, dug for clams, went surf fishing. They bought local produce grown by the Japanese farmers whose fields of bright red and yellow carnations sprawled over the sand hills.

In the prewar years, if you lived at the beach year-round (which was considered peculiar by most L.A. residents), chances were you worked there, too. There were glass and silk factories, small dairies and hog farms. I am sad I never knew Los Angeles in those early, uncrowded years when the only freeway was the beautifully landscaped Arroyo Seco Parkway in Pasadena, and the buildings, like the dresses of the day, were all curves and soft lines. I would have liked to enter Los Angeles along highways lined, not with subdivisions, but with the kind of roadside commerce that once marked the transition from country to city: produce stands, canary farms, chicken ranches, frogs for sale, as well as (it was L.A., after all) signs advertising psychic mediums, painless dentistry, and Chinese herb doctors.

World War II changed everything. Factories and oil refineries that sprang up to feed the war effort caused air pollution much worse than the smog we lived with in the 1960s and 1970s. On one long-remembered day, the city was crippled by air so thick that people fell ill, visibility was reduced to three city blocks, and planes were unable to land. With the influx of wartime workers, the sewer system became overloaded and raw sewage flowed into the bay, littering miles of sand with used condoms. Beaches were often quarantined.

Japanese farmers were shipped off to Manzanar, and their fields, which had long isolated the beach cities from Los Angeles, gradually filled with housing and industrial developments. Sand dunes were bulldozed so beach-goers could walk unobstructed from the shops and restaurants along the Strand to the water.

Construction of the Hyperion Treatment Plant eventually eliminated the sewage problem. Sand excavated from the plant site was distributed on beaches around the bay, extending them to unnatural widths; beachgoers now had a sole-burning two-hundred-yard dash to the water. The beach began to look less like a natural environment and more like a combined volleyball court and picnic ground. The jetty and breakwater built at Redondo's King Harbor created a pattern of erosion that ate away the beaches. To rebuild them, a pipe was extended two miles out to sea, spilling onto the beach a slurry of sand and seawater along with an occasional bizarre creature from the ocean floor.

The beaches weren't the only thing that changed. Even before malls sprang up in adjoining cities, Redondo Beach's downtown had disappeared—a casualty of efforts to modernize the pier area. Redondo joined the other beach cities in becoming a bedroom community, like a child's nursery, a place to play and sleep. More than suburbia, it was "surfurbia."

By the 1970s sleepy middle-class Redondo Beach was being seduced by the idea that it could regain some of its old glory. It would become a destination for well-heeled tourists and offer beachside living for all those swingers, divorcees, and others whose newfound freedom necessitated an ocean view.

Many of the grand turn-of-the-century houses overlooking the beach were inhabited by wealthy but thoroughly dotty old widows. My friends and I ran errands for some of these women, including one who always insisted we admire her latest cockroach kill, which she kept on display in a mayonnaise jar on the kitchen counter. Inevitably, when one of these old women died, the house was demolished to make way for a high-rise apartment or condo building.

The ornate but seedy Fox Theater next to the pier was razed, much to the relief of local parents. (The theater had been the exclusive domain of rowdy substance-enhanced teenagers—as well as the occasional hapless family of tourists—who sat with their feet on the seat backs in front of them to avoid the rats rummaging in the spilled popcorn.) The pier was rebuilt; it was cleaner, safer, and afforded ample parking. I missed the rickety old wooden structure with gaps between the boards that allowed an occasional glimpse of water beneath. Walking on the newly paved pier was no different from walking down the street.

I had grown up among people oriented to the west. Surfers scoping the waves, commuters checking to see if they should drive with the top down today—every morning we all looked to the ocean. Was it gray and choppy, blue, white-capped? The water told us what was happening in the sky—

whether to expect wind or overcast clouds, or if the fog was coming in. A smudge of brown on the horizon before eight A.M. meant there would be lung-searing smog levels by midafternoon.

On clear days Catalina Island materialized on the horizon and the remains of the *Dominator* (a freighter that ran onto the rocks in 1961) were visible off the Palos Verdes Peninsula. Walking, driving, or bicycling, all routes seemed to lead west; we might start out to do an errand on the Pacific Coast Highway and instead find ourselves pedaling along the Esplanade watching a spectacular smog-induced sunset. Even in our sleep, the sound of the foghorn drew us toward the west.

In my midteens I started resisting the westward yearning that had brought my ancestors here. Tired of the endless summer, I stopped going to the beach entirely. (My swimsuit tan lines remained for years, much to the amusement of college boyfriends who couldn't imagine me as a surfer.) I turned my back on the bay and looked inland, hoping for a glimpse of Mt. Baldy—reassurance that a natural world existed beyond Los Angeles. The beach had become nothing more than a backdrop for other people's fantasies, a sanitized, artificial version of the natural environment earlier generations had known.

The suspension of time that I had enjoyed as a child became oppressive to me as a teenager. Throughout the school year the sun shone, a lawnmower droned outside the classroom, the kids slumped over their desks. Everyone under thirty talked in vague, brief sentences, camouflaging any telltale signs of earnestness, ambition, or intelligence. (My monosyllabic surfer speech was punctuated by the long pauses of someone very stoned or, in my case, paralyzingly shy.)

L.A. culture encouraged us not to take anything too seriously. The values that our great-grandparents' generation had brought with them from the Midwest—moral rectitude, moderation, and hard work—had evaporated after decades of exposure to the California sun. Apparently, turn-of-the-century fears of the enervating effects of the climate had proved correct. The idea that working hard might "get them somewhere" made no sense to the young people of the South Bay; they were already here, where everyone else in the country wanted to be. Besides, success seemed to have little to do with ambition. It fell randomly on the Southland like meteorites or freakishly large hail; it seemed you could up your chances of being clobbered by fame and riches simply by spending every available minute outdoors.

Beach cities teens sometimes got tired of waiting and flung themselves randomly at life. It was not unusual to hear that a girl we knew had become a stripper in Tokyo or gone to Central America with a man twice her age. Others joined the Moonies or the Nishiren Shoshu Academy, whose meetings alternated mesmerizing, truly spooky chanting with rousing testimonials to the power of chant to bring you anything you wanted. (Need money? A new car? Chant for it!) Boys were not immune either, although their random acts often ended more violently: one gangling boy who used to beat me up in fifth grade was killed in a drunken drag race; another, my first boyfriend in junior high, died when he dived off the roof of a house and missed the pool.

A sign at the entrance to Redondo Union High School proclaims it the home of athletes and scholars, but instead of football heroes and Phi Beta Kappas, the school produced people more quintessentially Californian: the Smothers Brothers, Gerald Ford's would-be assassin Lynette "Squeaky" Fromm, and porn star Tracey Lords. Four years after I graduated, the writer Cameron Crowe went undercover as a student at what he called "Ridgemont High, Redondo Beach." The book (and later the movie) *Fast Times at Ridgemont High* was full of bored, disengaged teens distracting themselves with drugs, rock concerts, and inappropriate sexual partners. Apparently not much had changed, although there was one difference: the chronically stoned surfer Spicoli was referred to as the last of a dying breed—in my day, surfers ruled.

As a high school senior, I had no ambition other than to get out of L.A. Southern California was paper thin, insubstantial, all surface. I longed for mystery, complexity, depth. I wanted solidity. Mountains, not ocean. Buildings that were allowed to stand until they fell. I wanted people who weren't afraid to be passionate about something. I wanted something old and real. Clear air, silence, snow.

After an endless day of high school pettifoggery, I would come in and throw myself on the bed, relieved to be surrounded again by things that mattered. The walnut bureau next to my bed gave off a scent of lemon oil. The afternoon sun poured through the west window until it seemed the room couldn't hold it all. I grew drowsy breathing the warm air in and out and watching the dust motes float away with my breath. Released from the endless present of Southern California, I entered a world where things changed and grew and died. My grandmothers, the mountain stillness, L.A. sunshine, and Neil Young singing "After the Gold Rush" were all there in the room with me.

Two years later, at the age of nineteen, I left Southern California for good.

FIVE

Reunion

It was coming up on the hundred-year mark since Percival and Odalie Burns had established their homestead along Thornbush Creek, and Dooley was planning a huge family reunion to celebrate the occasion. He liked the idea of the ranch filled with friends and family members enjoying themselves together just like in the old days. There would be lots of food. Music. Some appropriate words said. "It's too bad nobody dances anymore," he said, remembering the old days in the schoolhouse.

Liz was less enthusiastic about the reunion. Unlike Dooley, who grew up with the bustle of a small community at Burns Ferry, Liz has lived most of her life in the company of one or two other people. She grew up in the lull between the canyon's heyday in the 1930s and its rediscovery by outdoor recreationists in the 1970s. Throughout most of her childhood, Burns Ferry was a ghost town. In the years since, most of the buildings have collapsed, burned, or been pulled down for the lumber.

In the summer the number of people on the river has increased so that the canyon echoes with the sound of boats, and there are people camping on the beaches, boaters in trouble, and tourists tramping the path around Ram's Head Rock to look at the petroglyphs. But there are times when the ranch returns to its former isolation. In the spring when the river is running too swift and high to cross safely, Liz may not leave the ranch for weeks at a time. Sometimes Dooley and Bertie are staying elsewhere, or they can't get out to the ranch because the road is washed out. There may not be a visitor for months.

Liz accepts such enforced solitude with equanimity. What is unsettling to her is an event such as the reunion, where she would be expected to be sociable with dozens of people she barely knows. Her flowers were bound to get trampled. There would be fires on the beach to keep an eye on. And she would have to ride herd on any kids roaming around unsupervised.

"Some kid's bound to get hurt or break something," she griped to me on the phone. "Why don't you come down for the barbecue so I don't have to face this crowd by myself." The weekend of the reunion, Bertie was in Portland playing a gig with the bluegrass band. Like a lot of hired hands, Bertie tended to disappear just when she was needed most.

The afternoon of the barbecue, the river road was lined with parked cars, and a jet boat waited to ferry people across to Burns Ferry. Taking a fast boat across the river was like driving a car over a bridge: suddenly we were there. I had always enjoyed the gradual transition from shore to shore afforded by the slow motion of the rowboat. The rhythmic lapping of the river, interrupted by the occasional awkward plash of a mistimed oar, washed away the sound of the car's engine still buzzing in my ears after the long drive.

We pulled into shore alongside several other boats. The white sand beach, usually a smooth expanse with a single trail of bootprints up to the house, was crosshatched with footprints. Loud honky-tonk music, talking, and laughter came from the yard hidden behind a grove of trees. The noise seemed out of place. On most days, the silence here is so deep that the scrape of the boat against the sand, the rattle of the anchor chain, or the cry of a killdeer as it rockets out of the dry grass, echoes like a bucket dropped in a well.

Behind the ranch house, the barbecue was in full swing. Dooley, carving thick slices from a whole roast pig, waved hello with his electric knife. I could tell by the extra little flourish he used to put the slice of meat on someone's plate that he was enjoying playing the host. Middle-aged women served up potato salad from huge tubs and handed out slices of pie on paper plates. Dozens of people, mostly young children and old folks, ate at picnic tables covered with checked tablecloths. Older kids perched on the rock wall surrounding the strawberry bed. The men sat in lawn chairs, leaning forward over plates held between their spread knees. A little girl ran by me, and a woman called after her, "Roxy, honey, what'd you do with Mama's beer?"

I spotted Liz leading Dixie, the gentlest of their three horses, around the yard. A small boy, holding the saddle horn like a joystick, was trying to steer the mare where he wanted it to go. Other children impatiently waited their turn for a ride. A young mother held a toddler who kept reaching out to the horse. Liz yelled at some kids to stay out of the flower bed.

Instead of her usual canvas coveralls and boots, Liz was wearing jeans, a new T-shirt, and tennis shoes. Her hair, often disheveled from the wind and the sweaty work of the ranch, was braided smooth. The harassed look left her face for a moment when she saw me approaching. "Weezer!" Bertie had started calling me Weezer (short for Louise), and Liz had recently picked up the habit, too. Ordinarily, I wouldn't have been able to stand the name, but here on the ranch, it seemed to fit. Liz glanced down—somewhere around the vicinity of my knees—and said, with a sigh of relief, "Glad you're here."

Liz rarely looks people directly in the face. When I first started coming to the ranch, I thought she avoided eye contact because I was an intruder and she was ignoring my presence. Now I know it is not that simple, as nothing about Liz is simple. Self-consciousness about her bad eye is part of it. There is also the rural culture's discouragement of open curiosity about anything, including strangers. When I mentioned it to a mutual friend, she told me, "You know, Gina also had that way of not looking when she was talking to somebody. May've been an Indian thing." Liz's mother had spent much time with the Nez Perce and spoke their language. Perhaps she had adopted their indirect manner of speech as well. The lack of eye contact with Liz bothered me for some time because it often appeared she wasn't listening. Eventually I discovered she is listening carefully all the time. Sometimes she seems to hear things you don't even say.

Talking to Liz, I discovered, is like talking to someone on a long car trip. There are extended silences, then periods of comfortable conversation. It is, in fact, easier to open up to someone who isn't looking directly at you. It's a relief not to be trying to make interesting, self-revelatory conversation across a coffeehouse table. Women friends so often want immediate intimacy. They share secrets before I even know basic stuff—their dog's name, when their birthday is, whether they like rum raisin or butter pecan. I find myself watching them watch me, gauging their reaction to my words; their faces blur and their words run by unheard as I try to formulate some witty thing to say next. Around Liz, I have to listen, really listen, because she never indicates with a glance that she is about to say something, and she tends to mumble.

"I don't know even half the folks here. They've been coming from all over—Grangeville, Orofino, Kamiah." She nodded toward a group of young men in cowboy boots and tight shirts standing by the gate, drinking beer. "Those guys camped on the beach last night. They were playing 'Pretty Woman' over and over for hours. We found one of them in Pop's junk pile, passed out."

I asked who they were and she shrugged. "Relatives on my mom's side. They're from up around Elk City." Elk City is a tiny logging town in the Clearwater Mountains about 120 miles from Lewiston. It's about as far away from anything as you can get. Beyond the town stretches the Selway-Bitterroot Wilderness, the largest wilderness area in the lower forty-eight.

A woman came up to Liz and asked her to get something from the kitchen. Liz handed the horse's lead to me. I held Dixie still while the toddler patted her nose with a flat palm, his mother repeating, "Hor-sey, hor-sey." A seven- or eight-year-old girl climbed onto the horse. She said her name was Bandy. I didn't have to ask how she got her name: her skinny crooked legs in cutoff shorts barely reached across Dixie's wide back. She yelled to a woman at one of the picnic tables, "Grandma, take a picture of me!"

I had been reluctant to come to the reunion, worried that I might be intruding on a private family gathering. But it was obvious no one recognized I was a stranger. A man walked up, scratched the horse's ears, and asked if I had seen Uncle Ralph. I smiled and told him no and suddenly felt perfectly comfortable. There were other people here who were friends or neighbors of the Burns family: Dooley's friend Lon; Dorrie McAlpine, who used to work at a lodge in Hells Canyon when Dooley was running the mail route. I recognized the rollicking speech of a local cowboy poet and turned to see him greeting a friend with a voice as big as his belt buckle.

The Burnses apparently had a looser definition of family than the Freemans did. For us, family had been narrowly defined: in-laws were accepted only reluctantly; stepchildren and adoptees were second-class citizens; black sheep were banished from the family. Friends, co-workers, and neighbors were beyond the pale. No wonder our family gatherings were small. I had never been to a family reunion large enough to include people I didn't know. A childhood friend, struggling to explain why she had felt uncomfortable at our house, told me once, "The Freemans were like a club no one else belonged to."

A horse-crazy teenager, who had been hanging around watching the younger kids, asked if she could ride the horse. She kept assuring me she "knows all about horses," but then couldn't figure out how to put on the bridle. I was no help. Liz and Bertie had shown me a dozen times how to buckle the leather straps around a horse's head, but all I could remember was the metal part goes in the mouth and the dangling straps, somehow, impossibly, fit around the neck and ears. Liz came up and took the bridle out of the girl's hand. She removed the bit gently from the horse's mouth and put it in correctly. The girl rode off cautiously, and Liz watched her go. "She lives in town," Liz said, as if that explained everything, and turned to go into the house. She wanted to show me a new graphic arts program she had just gotten for her computer.

Liz was still muttering about a comment made by one of her cousins, who had once lived in the house, about how well Liz had "kept up the place." The house had stood empty for several years until she and her parents moved in. "You should have seen it," she said, "Every window was broken. There were bats living here. It stunk to high heaven—a raccoon had died inside the walls. There was only one light bulb in each room, hanging from a wire. The walls were just bare boards. Pop and I wired the whole house for electricity and put up insulation. My mom planted all these flowers."

Though Liz has made other improvements since then, the house is far from luxurious. The supply of running water, brought from a spring high up on the hill, is often erratic, freezing up in winter and running low in the summer. The smoke-darkened particleboard walls are covered with old quilts, Indian blankets, Mexican serapes, Chinese shawls, paisley Indian curtains, and crocheted bedspreads that provide a layer of insulation. Throughout all but the hottest part of the summer, the living room stays cool and musty as a cave.

On winter nights, the firelight flickers and flares, revealing skins and animal faces in the shadows, a bearskin, a bobcat, a cougar, a badger with long curved claws. Rattlesnake skins hung on a nail move with a dry rustle whenever the door bangs open and another load of wood is brought in. From upstairs comes the murmur of doves as they settle down for the night in their cage. A crack like gunshot sounds across the river, followed by a long rumbling; the winter rains have brought another rock slide down Prospect Ridge.

Even in midsummer the room was dim, the windows obscured by ivy clinging to the porch. There was just enough light to make out the titles of books lining the east wall: books on horse training, how to make your own moccasins, *The Boy Mechanic* and the *I Ching*, the art of Rembrandt and Raphael, and a whole shelf of Star Trek paperbacks. From the ceiling hang Indian drums made of deerhide tanned by Bertie. On the walls hang Liz's oil paintings, Hiram's fiddle, a rapier, a butterfly made of balsawood, an Indian flute in a beaded case, a battered bugle, a long twisted piece of driftwood, a rawhide whip, an old ship's lantern. On top of the piano is a doll-sized schoolhouse and a canning jar full of turkey feathers.

The computer sits within fanny-warming distance of the wood stove, cold and smelling faintly of ashes at this time of year. Liz showed me some stunning graphic art she had done on the computer—delicate, fluid images of horses, dragons, and fish.

From the beach came the sound of fireworks. Liz said, "It's those guys again. They bought a bunch of fireworks in Kamiah. I hope they don't start a grass fire." Kamiah, about sixty miles from Lewiston, is at the heart of the Nez Perce Indian Reservation. Along the winding highway that follows the Clearwater River, roadside stands sell firecrackers, tax-free cigarettes, beaded key chains, and other Indian knickknacks.

Liz went down to the beach to check on the crowd shooting off fireworks. Outside in the yard, the hubbub was dying down as people wandered off. They were probably visiting their favorite spots—schoolhouse, orchard, garden, barn. Young kids might be exploring the empty chicken house that smells of old feathers, or wandering up the dirt road, picking blackberries and looking for arrowheads. It was mostly old folks left at the picnic tables, sifting through piles of loose pictures and looking at photo albums.

I sat down and looked at the photographs in front of me: community picnics along the creek in the 1920s, the girls in white dresses and Dutch boy haircuts, the boys in overalls; good-looking young cowboys in sheepskin chaps posed on the steps of the dance hall, cocky as all get out; the man called Silver Buckle with a nine-foot sturgeon draped over the back of a mule, its tail dragging the ground. Next to me, two old women couldn't agree on whether the woman in an unlabeled photo was Aunt Gert or Aunt Lacey. A large framed photo—Dooley must have taken it down from his wall—of Percival was in the place of honor in the center of the table. He must have had blue eyes or green, like Liz, for his eyes in the black-and-white photo are pale and ghostlike. With his long beard and pipe, he looks like a born storyteller and good company to sit with by a fire on a winter night.

In the 1880s Percival and his wife, Odalie, homesteaded land just east of Steptoe Butte, a bare knob that sticks up above the surrounding hills of the Palouse in what is now eastern Washington. There he planted grain and started a hog-raising business. Unfortunately, Percival's hired hand managed to get involved in a "horse-stealing scrape" and went to jail. Percival generously put up a bond of two thousand dollars for the man, who immediately jumped bail and disappeared from the area. The family lost the farm.

Percival was so disgusted, Dooley said, that "Grandpap changed his attitude towards life and said that never again would he gather a coin. Raise enough to eat and that was it. And that's what he did. He hunted and fished and raised gardens and played around. And he built things." The place he chose for his retreat from the world was this rugged spot in the Snake River Canyon, eighty miles south of Steptoe Butte.

I've heard this story more than once, and Dooley always uses the same

Sheep grazing near Lewiston, 1904. Nez Perce County Historical Society, Lewiston, Idaho. 89-27-7.

odd turn of phrase, "Nevermore would he gather a coin." I have no doubt these are the words Dooley heard from Percival himself when he was an old man with a white beard flowing down to his chest.

Percival and Odalie settled on a wide bar on the Idaho side of the river, upriver from Lewiston. In the steep, arid Snake River Canyon, there weren't many good homestead sites. "In them days," Dooley said, "a prime thing was a crick where there was an adequate supply of water. Then it was possible to raise up enough stuff to feed a family or two or three families. That was the case at Burns Ferry." Thornbush Creek tumbles out of a side canyon that leads up to the top of Craig Mountain. The water is still cold and clear even in midsummer. A broad alluvial fan at the mouth of the creek provided enough flat land for several families and their gardens, outbuildings, and pastures. Before the homesteaders arrived, the land along the creek had been the site of the Nez Perce village of Hetéwisnime.

Dooley tapped a photo of a ferry full of sheep. "Now, that was some trick. Gettin' the sheep onto the boat." As a young man, he had worked for a big sheep rancher up in Hells Canyon; his least favorite memory was of trying to herd the sheep onto the ferry to get them across the river and on up to their summer pasture in the Bitterroot Mountains. There are no bridges on the middle Snake above Lewiston; canyon residents have always

Ferry loaded with sheep, Hells Canyon, 1943. Photo by Kyle Laughlin. 99-G-006-08. Historical Photograph Collection, University of Idaho Library, Moscow, Idaho.

had to rely on boats—Nez Perce bull boats, cable ferries, jet boats—to get themselves, their freight, and their stock across the river.

When the Burns family settled along Thornbush Creek, the road into the canyon extended only to the steep bluffs on the Washington side of the river, directly across from their homestead. To continue on into Hells Canyon, homesteaders and miners had to pack their supplies by mule over steep and treacherous trails along the breaks. The other option was the steamer, which made the trip upriver only when there was enough freight—typically, head-high stacks of three-hundred-pound bags of wool—to warrant the risk of taking an unwieldy paddle wheeler through the rapids. There were months on end when the steamer couldn't make it upriver because the water was too low. With his blend of the visionary and the practical that epitomizes the Burns family, Percival saw that his homestead would be an excellent location for a ferry.

Despite his insistence on not bowing to the almighty dollar, Percival's ferry was a profitable enterprise, at least in the summer months. The ferry was inoperable in winter when the river froze, so Percival had plenty of

Waiting for the steamer at Imnaha Landing, ca. 1905. Photo by Henry Fair. 99-13-8. Nez Perce
County Historical Society, Lewiston, Idaho.

time to work on his improvement projects. He built a schoolhouse, a
dance hall, a post office, and a fifteen-foot waterwheel that powered a
gristmill and generated enough electricity for twenty-five-watt bulbs in
all the buildings. People came from miles around to marvel at the bright
lights. (The waterwheel froze up in winter, and the creek residents had to go
back to using coal oil lamps.) For the kids, Percival built a water-powered
merry-go-round painted with green and red stripes.

Some of Percival's projects backfired. Believing, as most of the settlers did,
that cool well water was more sanitary than creek water, Percival dug a well
behind their house. Mosquitoes bred in the standing water and infected the
family with typhoid fever. Their oldest boy died, and ten-year-old Hiram
(Dooley's father) lost an eye from the fever.

The homesteaders, most of whom were from the South and Midwest,
initially had tried to recreate the type of farm they had known at home:
fields of wheat, corn, and oats; a large garden; and a few chickens, hogs,

and cows. But as in other places in the arid West, homesteaders could not make a living on the 160 acres that would do for a farm in the well-watered Midwest.

Dooley said, "At that time, why, there was people fluctuating all over. Every hundred and sixty acres that was worth a damn, there was somebody trying to live on with a family. They would prepare everything they could in the winter so the wife and kids could survive through the summer and raise a garden and do whatnot. Then they went somewhere to try to make a dollar or two, like out in the harvest, or a lot of them went to work for the Forest Service. Built a lot of trails through the mountains. The big bugaboo all during the homestead days was the taxes—raising that cash to pay the taxes was a real hard thing to come by. An awful lot of them lost their property for taxes."

By the end of World War I, fewer than 5 percent of the original homesteaders remained on their claims in Hells Canyon. Those who survived did so by buying up abandoned claims until they had enough acreage to run a large herd of cattle or sheep. The ever-resourceful Burns family gradually extended their holdings; from the original quarter section, their holdings grew to include over five miles of riverfront as well as a timbered portion on Craig Mountain.

In a turn-of-the-century photo a bent old woman wades through the muck of a cattle corral in an ankle-length print dress and white apron. In the corner of the picture a cow stares directly at the camera. The woman is carrying a feed bucket. Her toothless mouth is just visible beneath the oversized brim of her ruffled bonnet. She looks grim and weary.

In a photo taken ninety years ago in the same yard where the barbecue was being held, the trees are small, the light harsh. Now the trees are taller than the two-story house, and even on that warm summer day the yard was cool and shady. The trees are a fast-growing variety of locust called "trees of heaven." With their long dagger-shaped leaves and trunks the color and texture of elephant skin, they look ancient and exotic enough to shade a Hindu god with upturned palms and a foot cocked in midair.

Hiram's wife, Louetta Weilmuenster, was born at Gap Creek, five miles downriver from Ram's Head Rock. The Weilmuensters, one of the earliest pioneer families in the area, had established a homestead on the large bar where Gap Creek flows out of a slot canyon. As an old woman she remembered being terrified by the Chinese gold miners who came upriver rowing long open boats with dragonhead bows. Louetta ran and hid whenever she heard the rhythmic, singsong call of the oarsman keeping the men rowing in time; she had been told the Chinamen would steal her. By the time she grew up and married her neighbor, Hiram Burns, the Chinese were all gone—chased out, poisoned with strychnine, shot and dumped in the river.

Percival, who loved his fiddle, taught all his children to play music. Hiram grew up to be a popular musician at local dances; he was known as "the best darned one-eyed fiddler on the Snake River." A picture taken in 1914 shows the Burns family dressed in their finest. Three of the men hold fiddles, and a boy, wearing a suit and vest and holding a shepherd's crook, stands beside a sheep with a ridiculously thick coat of wool. A young girl, hands clasped at her waist, smiles self-consciously in her dark gingham dress with cap oversleeves and a single precious string of beads.

Residents of the Snake River Canyon took the Great Depression in stride. Farmers and ranchers throughout the country had overextended themselves during the war years, and the 1920s had been a difficult time for them. The 1930s felt like more of the same, but with the difference that now everyone else was in the same boat. In fact, rural residents were better off than many city residents, and remote locations like the canyon became refuges for people unable to make a living on the outside. Schoolteachers who couldn't find a job in the city and young men who came looking for work on one of the sheep ranches were able to find a place in the canyon community.

Families with many children to feed settled on a few acres of land, grew their own fruit and vegetables, and kept milk cows, chickens, hogs, and sheep; every week they went to town with cans of cream or a case of eggs to be sold or traded for meat at the slaughterhouse. Dooley said, "During the Depression, when almost everyone was hungry, we had plenty."

There was a spirit of cooperation among the families at Burns Ferry and the other residents—fewer than three hundred people—strung along the Snake. Old Jim Chatham taught the delicate art of shearing sheep. Dell Russell, who eked out a living placer-mining for gold along the river, showed anyone brave enough how to get honey from wild beehives. L. O. Bosley from the mountains of West Virginia demonstrated the basics of making moonshine. Cora Fountain shared her secret of using a filter of burnt wood shavings to give white lightning a nice gold color like store-bought whiskey.

Isolated and cash poor, the residents of the canyon depended on each other for entertainment. In the winter, they often gathered at Burns Ferry for all-night dances. After the dance hall burned down in the early 1930s, dances were held in the schoolhouse. Arriving by boat, mule, and horse from ranches up and down the river, the partygoers washed up and changed into clean clothes. They pushed back the desks in the one-room schoolhouse and cleared a space in a corner for the musicians: Louetta on piano, Hiram on fiddle, and Uncle Bill on banjo.

At a time when people in the city were dancing the rumba, the canyon folk danced the two-step, the circle waltz, and the quadrille. The small schoolroom, packed with fifty or more bodies, grew warm and stuffy. Children nodded off in corners, sleeping on piles of coats. Men stepped out into the cold night air for a nip of moonshine. The women laid their babies down in the narrow bed in the tiny teacherage at the back of the school.

Cowboys with a few too many nips under their belt fought over who was going to dance next with the teacher, sixteen-year-old Miss Winaford Chance. If they were really drunk, they engaged in a favorite game: cutting off each other's necktie at the throat with a quick slice of a knife. When the dancers were tired, the musicians wanted a break, and the cowboys needed coffee, the women laid out a midnight supper of chicken, turkey, and sandwiches. Cakes brought on horseback over the mountains were frosted to cover the cracks.

After dinner, any strangers in the crowd were called upon to tell a bit about themselves. The ranch community, ever wary of outsiders, liked to put these people to the test. They insisted that the stranger—man or woman—tell a story, recite a poem, or sing a song. If they refused, they got a hot skillet on the backside. More than one fight was started when some stubborn cowboy refused to sing.

In one of the yellowed scrapbooks being passed around was a newspaper clipping headed "Death of Pioneer Lady." The obituary mentions Odalie's "big-hearted ways and kindly disposition to all people." After his wife died, Percival built a house high on Craig Mountain. He planted a small garden and kept a few chickens. He fished and hunted for grouse. To bring in a little money to buy staples such as salt and tobacco, he made moonshine.

Many of the settlers who moved into the rough backcountry were from the states of Missouri, Arkansas, and Tennessee. They were accustomed to making their own wine out of whatever was handy—chokecherries, rhubarb, or dandelions. But Prohibition added a whole new incentive to the liquor-making process: canyon residents could get as much as ten dollars for a gallon of moonshine. With a well-placed still hidden in one of the side canyons, a moonshiner could escape detection for years. And there was plenty of time to build up a good business because the statewide "dry law" wasn't repealed until 1935, two years after federal prohibition ended.

The photo that seemed to be everyone's favorite is of Percival and the Burns boys getting ready to go to town. A keg of moonshine is strapped to the back of a horse. A grinning teenager holds a crock jug over his shoulder with one finger. Percival, looking threatening, holds a .38 Colt Special on his knee. They all seem to be taking a rakish delight in their status as outlaws.

The picture reminded me again how "other" I am here. My great-grandfather was a strait-laced teetotaler with pince-nez glasses. Liz's great-grandfather looked like one of the feuding Hatfields and McCoys. At the time the Burnses were cooking up moonshine, my grandmother was working for the state liquor control board. Just as the Burnses relished their outlaw role, Smokey enjoyed the part she played in sting operations targeting bars that served drinks after hours. In a game of cops and robbers, there was no doubt that my family and Liz's would have been on opposite sides.

My father described the scene to us just the way his mother used to when she came home from Sacramento to visit. Smokey would walk into a speakeasy, sit down, order a drink, and slowly take a sip. If she tasted hard liquor, she would open her silver cigarette case, light a match, touch the end to a cigarette. Then she would raise her arm and wave the match in the air with a dramatic flourish. The officers waiting outside for the signal would come rushing in to make the bust. I imagined my father, a boy in white shorts with suspenders and patent leather shoes, listening wide-eyed to the tall, dark-haired woman—a stranger with bright red lipstick who was always swearing or laughing or doing both at the same time.

The divide between the Burnses and my family seemed immense. I remembered the finality with which Liz dismissed her young relative as a know-nothing because she lived in town. The members of my family were proud of being modern, well-educated city people who had shaken the dirt off their heels long ago. (In the 1940s my father dated several girls from Dust Bowl families precisely because his mother didn't want him associating with "Okies.") They believed their children were not suited to a country life.

When Smokey was about to get remarried and move to Alaska, her father wrote to dissuade her, saying, "You don't belong in Alaska. You're a city girl." She was then in her early fifties. My mother used the same line on me when, as a teenager, I talked about taking an Outward Bound course, learning rock climbing, or moving to Colorado. "You're a city girl," she said. "What makes you think you can do those kinds of things?" It's been almost twenty years since I left L.A., but in Idaho I still feel like I'm wearing a big sign that says, "City Girl, Kick Here."

Dooley sat at one of the tables, signing photos of the *Caroline*. He has spent a lifetime ranching and was a riverboat pilot for only twenty-odd years. But with his boating cap and gold-rimmed teeth, he looks more like a ship's captain than a retired rancher. The old riverboats that Dooley used to run up the rivers have been replaced by faster but less elegant and much noisier jet boats. The even older paddle wheel steamers he remembers as a kid have been long gone from the Snake. His affection for the steamboats led him to buy the hulking, rusted *Caroline*, which had languished for years after a local restoration project (originally launched by Gina) was abandoned for lack of support.

The boat is now moored alongside the town park in Asotin. Undeterred by those who consider the faded *Caroline* an eyesore, Dooley can often be found on board, painting the railing or fixing a rotted staircase. He is often accompanied by his friend Lon, who has also invested time and money on the project. Together they plan the *Caroline*'s future as a place where weddings and senior proms will be held and schoolchildren can learn about the history of navigation on the Snake River.

Dooley was born at Burns Ferry in 1915 and did all eight years of his schooling at the small schoolhouse where he now lives. As a child, Dooley was surrounded by storytellers: Indians like old Joe Albert who told Nez Perce stories about the battle between Ant and Yellow Jacket and how Coyote lost his eyes and transformed himself into an old grandmother to get them back again; and his uncle Bill, who told prophetic tales of a population explosion, a race war, and an out-of-control national debt.

Old-timers gave exaggerated accounts of their stake in the Gold Rush of 1863. Others described in lurid detail how they found the mutilated bodies of ten Chinese miners in the river, floating down from Deep Creek, where they had been massacred by whites who were after their gold. Many men could tell of the wreck of the steamer *Imnaha* in 1903 and of the famous Brownlee shooting near Crooks Corral the following year. This story was a popular one for retelling because it had a particularly satisfying ending: Brownlee's killer got what was coming to him. A mob of townspeople from Whitebird took him from the sheriff's buggy at gunpoint and hanged him from a tree.

I noticed that among all the pictures on the table, not one included a Nez Perce Indian, although they had been a regular part of the settlers' life at Burns Ferry. I would have liked to see a picture of Joe Albert because Dooley once told me he was "the oldest-lookin' thing in creation you ever saw." Joe Albert was a favorite of the children because he liked to tell them stories. Dooley said, "He had wrinkles everywhere, deep wrinkles on his hands, face. He used to measure us kids' feet by counting the wrinkles on his fingers." He had held up his hand, misshapen from arthritis, to show me how the old man would measure a child's foot against his fingers, the wrinkles like marks on a ruler. "Then he'd come back the next spring with moccasins he'd made for us."

I browsed through some more recent photos of the Burnses from the 1940s and 1950s. There was a look about them I had seen in other photos of canyon residents of the period: proud, happy, confident. They didn't show the exhaustion of the original homesteaders, who had been just hanging on by the skin of their teeth. With hard work and a good business sense, they had built on the groundwork laid at such expense by their grandparents. Others had quit, given up, been forced out, but these were the ones who had stuck it out. They had made it. Their homes were comfortable, their land was productive. They were able to provide themselves with almost everything they needed. There must have been great satisfaction in that.

Change came to the canyon with America's entry into World War II. Many people left to work in the factories and shipyards on the coast or down at the processing plants in southern Idaho, which were in operation around the clock to meet the demand for dehydrated potatoes for the troops. Hiram had to shut down his grain warehouse because jute to make sacks became scarce; farmers switched to storing wheat in elevators and shipping it by rail.

People were buying clothing made with the new synthetic fibers developed in wartime laboratories. Vets home from the war were eating differently. Eating mutton was like wearing long underwear or red suspenders: good enough for back home, but out of place in the big city. With the decline in demand for wool and mutton, the sheep-ranching business took a nosedive.

Ranchers in the lower end of the canyon, such as the Burnses, were able to switch to cattle because the layers of basalt in the canyon walls stop rainwater from draining away, funneling it to the breaks, where it emerges as springs. Higher up in Hells Canyon, the basalt in the canyon walls is more jumbled, allowing what little water there is to seep down through the soil and disappear. The only animals it was practical to run there were sheep, which can survive on less water than cows. When the market for sheep dried up, so did the ranches.

Electricity and phone lines finally reached the canyon in the 1950s. Ironically, residents became more closely linked to the outside world just at the time they were becoming more isolated from each other. As more ranchers had sold their stock and land, the remaining ranches had grown larger. The distance between neighbors grew as the number of people in the canyon shrank.

The Burnses' land had plenty of water for cows, but they discovered there wasn't much money in running cows, either. By the mid-1960s Dooley had run through a laundry list of odd jobs trying to make ends meet. He finally took a job running the mail route by boat to ranches and small mining outfits in Hells Canyon. The once-a-week run took two days. In the winter, he was often the only person these ranch caretakers and miners saw for months on end.

Picking up a piece of pie, I sat down at one of the picnic tables and introduced myself as a friend of Liz's. The men and women nodded politely and went back to catching each other up on family news and asking whatever happened to so-and-so. I ate my peach pie and looked around at the people at the table—the little boy, face smeared with blackberry juice, the woman smoking a cigarette between bites of cake—and realized I envied them.

It had been just four months since my father died. My siblings and I no longer had a place to return to in the town where we grew up, on the edge of the bay where our ancestors had lived for over a hundred years. I envied the Burns family's connection to the land, the past, their family history. Although I have probably spent more time at the ranch in the last few years than any of these people, they have deep memories I will never have. Many of them lived here at one time or another, then left to live an easier life in town. Even those who never lived on the ranch had visited as children, swimming in the river and picking apricots. The men who have hunted and fished here know the mountain and the river in a way I cannot.

I realized that the reunion looked so peaceful partly because the personality conflicts and rivalries that exist in any family were invisible to me. But it also seemed likely that the emphasis on the ranch itself diffused those personal differences. My relatives' concept of family lay wholly within the difficult, ephemeral arena of personal relationships; membership was determined by whether an individual was in or out of favor in the current family circle. The Burns family had connections not only to each other but also to a third thing: the land. That connection seemed to me a much more enduring one.

Before the reunion broke up, the family gathered for a group photo. Everybody had to be in the picture: old women who didn't want to stir from their comfortable spot in the shade, babies who kept trying to crawl off across the grass, even me and the horse. Lon, Dooley's friend and partner on the *Caroline*, was jollying everyone into position. He looked through the camera and motioned me to move the horse to the right. "The horse's ass ain't in the picture." Several men immediately called out, "That's because he's behind the camera!" There was a roar of laughter. The camera clicked. We held still for one more, then "one last shot." The reunion was over. People began gathering up dishes, soiled tablecloths, and tired kids.

Later I was to look back on this easy, jolly scene and compare it to the tension-filled scene at the Freeman family reunion. With my siblings scattered around the country—Texas, California, Idaho—we chose Utah as a central meeting place. We camped at the Grand Canyon so there would be plenty of outdoor activities for the kids. Between us we had sixteen children, eight of whom had been born since our mother died, and another just weeks after our father's death. It was the first time we had been together since his funeral the year before.

For eleven years—from the time of my mother's cancer diagnosis to my father's death of complications from a liver transplant—our family communications had centered around the latest emergency. In long-distance phone conversations we had weighed the opinions of various specialists and the pros and cons of different treatment options. We had become skilled at negotiating with doctors and nurses and in dealing with insurance paperwork and hospital bureaucracies. In just the first year of my father's illness he had been in five hospitals in three different states.

The nurses often commented on how well we worked together, unlike many families they saw in similar circumstances. It was something we had been trained to do: all those times we had lined up behind a stranger's car waiting for my father's signal to push the stuck vehicle out of the sand. Grandma Smokey called us a tribe; our father called us his troops. Whatever we were, we were capable of acting together only under the direction of a leader. But Dad was no longer there to call everyone to order with an ear-piercing whistle. Mom wasn't there to lay out the schedule for the day like a well-drawn battle plan. Without our parents in charge of the reunion, we took forever to reach decisions about what to do and see—even what to cook for dinner. At the Grand Canyon Lodge we spent most of the time trying to locate family members who had wandered off or waiting for others to emerge from the restroom or gift shop.

Between our mother's death and our father's eight years later, our stepmother, our godmother, our only uncle, and our two surviving grandparents passed away as well. One after another, these huge elemental presences in our lives had fallen, the ground shaking under us each time, until when the air finally cleared, my brothers and sisters and I were amazed to find ourselves still standing. Still in our thirties, we suddenly had become the oldest surviving generation in our family. As baby boomers, we were used to being accompanied through every stage of life by an army of our peers. But few people we knew had lost one parent, let alone both.

No one at the Freeman reunion remembered to organize a family photo shoot until it was almost too late; some people needed to head out first thing the next morning. We lined everyone up in the dark, positioning them by flashlight. In the photos just our faces are visible, floating white against the dark night. We look scared, like untried troops pushed to the front of the charge.

Those members of the Burns family with boats started ferrying the others across the river. Dooley decided to go into town, and somebody offered him a ride. Finally everyone was gone. Liz let the cow dogs out of their pen. The cat, hiding all day, appeared again, hungry and complaining. We carried dirty casserole dishes into the kitchen, and I washed dishes while Liz wrapped leftovers in tinfoil and put them in the old-fashioned, waist-high refrigerator, painted a bright red to match the stenciled chickens on the cupboard doors. We didn't talk much, enjoying the blessed quiet. As we dried the dishes, Liz mentioned that as a distant relative was leaving, he slipped her his phone number. "If you ever get lonesome out here all by yourself, give me a call," he had said. "Sure I will," she laughed, "when the devil shows up on my doorstep wearing a fur coat."

The next morning, I unchained the rowboat and watched the shifting pattern of eddies on the surface of the Snake as Liz bailed water out of the bottom of the boat. She stood knee deep in the river, her long braid swinging with the rhythm of the bailing bucket. The only sound was the wind lifting off the river.

The Burns family reunion had left me with a mental scrapbook of the old days at Burns Ferry. Burdened with a useless nostalgia for a Southern California that no longer exists, I was relieved to find that I could have an interest in the ranch's past without regretting that it has faded away. I see only what is here now—the crumbling cabins and rusted tractors, the empty land, the silence—and it is enough. I looked down and saw a small pocketknife sticking out of the sand. I brushed it off and showed it to Liz. Turning it over in her hand, she said, "One of my cousin's kids lost this yesterday." Smiling, she slipped it into her pocket. "He'll be glad to get it back."

SIX

Family Traces

Every American family looks to one ancestor as progenitor of their family line: in families of European descent, such as mine, it is usually the first person to have crossed the Atlantic. But westerners have a second "family father": the pioneer who headed west. You can't be a true westerner without a family story about the original pioneer coming west for a new start. And the family history begins right there, with that person (usually male). Before that is prehistory, water under the bridge. What matters is when the family shook the dirt off their boots and headed west, and where they landed when they did. Everything else was B.C.W.: Before Coming West. I asked Dooley once where his ancestors came from originally, and he said with an apparent lack of interest, "I don't know. Fell out of the sky like the rest of us, I suppose."

Most of the people in California are relatively recent arrivals, having settled in the state since World War II. Their stories of coming west (or, increasingly, north or east) to California don't have the desert varnish of age of the pioneers' tales. In many older California families such as ours, the coming-to-the-West story has been lost as waves of emigrants flooded the state, one boom following another; the past has simply been washed away. No one in California bothered to keep track of such things as who arrived first: we were all from somewhere else and it didn't really matter where you came from and when, or even who you had been in that previous life. Everyone got a new start here. Without those coming-to-the-West stories, we may not think of ourselves as westerners, but merely as Californians, each person's life an individual, not a collective, tale. To reclaim some sense of myself as a westerner, it was necessary to reconstruct the story of how my family came west.

The story of how Percival traveled the Oregon Trail and established the homestead on the Snake River has always been carefully nurtured by the

Burns family. In our family, the stories of coming west have been allowed to atrophy. Instead of fully outlined characters like Percival, with his disdain for money and penchant for fanciful building projects, my ancestors were described with a mere word or a phrase.

My Swedish great-grandmother was "tiny"—I imagined her as a Scandinavian doll with rosy cheeks and red felt clothing. My great-grandfather was a "bounder," his sister a "harelip." One of my great-grandmothers was referred to as "a strait-laced Yankee Puritan" and the other as a "spoiled Southern belle." My great-grandfather's second wife was a "gold digger." Two great-uncles were given the damning label of "effeminate." Another great-uncle—one of eleven children—was remembered as "the smartest of his generation." One relative was noted for his habit of combing out his walrus mustache with a dinner fork at the end of every meal.

Often the only thing I heard about my ancestors was the way they died: "drowned in four inches of water," "trampled to death by his own horses," "died in childbirth because she refused to have her baby in a hospital." My grandfather's sister died of diphtheria at the age of five; her death entered the family lore because she used to wash her hands over and over, saying, "I must have clean hands when I go to Jesus."

It bothered me that so many of my ancestors had been reduced to a single, often grotesque image, no more than shrunken heads dangling from our family tree. I wanted to flesh out my ancestors and see them as real people. I began researching our genealogy, obtaining census and military service records and copies of marriage and death certificates. A history of the Swedish American Patriotic League in San Francisco allowed me a glimpse of my great-grandfather: Emil Hoberg, the league president, was described as "a tall man sporting a trim goatee, who was fond of reciting poetry." In a town history from New Hampshire, I read that Colonel Chase Taylor was a "man of great physical strength and energy of character," whose thigh was broken by a musket ball during a battle in the Revolutionary War, and he didn't realize it until "he attempted to rise and found his leg up in his face."

It took some investigating to determine who were the "pioneers" in my family. James Sr. and Agnes Inglis's journey from Scotland to America was documented, yet their arrival in the West was not. All I know is that by 1881 they were farming a piece of land outside Petaluma, near San Francisco. After James Sr.'s death, Agnes moved south to Florence, near Los Angeles, where her sons had put down twenty dollars on twenty acres of land. (With the Scotch laconism and lack of sentimentality that remained in the Inglis family for generations, her son noted the death in his diary: "Father died twenty-five minutes too one. Funeral Expences $71.35.")

I suspect that living in St. Joseph, Missouri, the jumping-off point for the Oregon Trail, had something to do with the Burns family's decision to head west. Having heard a great deal of talk about the territories, perhaps they finally decided to go see for themselves what the fuss was all about. Even though both Liz's family and my family probably originated from the same English-Scottish stock and had, in fact, settled only about 150 miles apart in the American Midwest, our families had little in common. The Burns family included farmers of "no particler religion" who prided themselves on being friends with Jesse James, the most famous resident of St. Joe; they chose to head for one of the most isolated and least-settled parts of the West. The Inglis family were city folk, strict Scottish Congregationalists who placed a high value on education and community standing; they wanted to move to a civilized part of the Far West.

Pioneer stories like the Burnses' emphasize the hardships endured on the trail and the backbreaking labor involved in proving up on their homestead. Such stories appeal to us because the pioneers are so quintessentially American: brave, energetic, optimistic. Homesteading (never mind the Indians who were there first) also seems wonderfully American: almost anyone could become a property owner if they put up a little money and worked hard to improve the land. At least in the rural West, pioneers are still seen as the good guys, despite efforts of recent western historians to dislodge them from their pedestal. A brave young hero, an arduous journey, success in an alien land: with all these elements, it's no wonder the Burns family story survived.

My family's story wouldn't make as good a tale to tell the grandchildren by the fireside. For one thing, the Americans weren't the good guys in early California history. The forty-niners didn't battle the resident Indians, they slaughtered them for sport and to clear the way for mining claims. No homesteading here; the new Californians got the land away from the Mexican ranchers by contesting their land claims in court. The Americans won either way: a rancher might be unable to prove his claim owing to the Spanish government's lax method of awarding land grants, or he would be bankrupted by court costs and have to sell his land. Americans bought the old rancheros, subdivided them, named their new "towns," and then went to the Midwest looking for suckers to buy the lots.

Iowa was a particularly good trolling ground, and undoubtedly James Sr. and Agnes Inglis heard one of these California promoters or read the broadsides distributed throughout the state. The railroads that served Southern California were engaged in a price war, and the cost of a ticket from the Midwest to Los Angeles fell from one hundred dollars to as low as one dollar. Special "emigrant trains" were outfitted to carry whole families west, lured by the promise of a healthy climate, cheap land, and high wages.

The Inglis family came west, then, in the relative comfort of an emigrant train, with seats that folded down into sleeping bunks and braziers to heat water for a cup of tea. This was not a high-adventure mode of travel (although they likely saw buffalo from the train and perhaps caught a titillating glimpse of the sinfully polygamous Mormons in Utah), and nobody bothered to tell stories about it to the grandkids later. Members of the Inglis family also may not have liked to admit they felt like chumps once they arrived. Many of these new settlers discovered that the reality did not quite live up to their expectations.

James Jr. had come west at the height of Southern California's first land boom, joining his brothers and mother in Florence. By the time his wife and five children arrived in 1889, however, the balloon had burst. A thousand people a month were streaming out of Southern California. There was no construction work. James had traded their property in Osage Mission, Kansas—four houses on a quarter of a city block—for a house and land near the town of Florence. His wife, Mattie, was shocked to discover that their new home was a "shabby two-room house," no doubt infested with the pernicious fleas mentioned in many settlers' letters home. She spent the short time remaining in her life pining for her home in Kansas. When Mattie died, nine months after arriving, James must have wished they had never come to California.

Like the Inglis family, the other branches of my family—the Johnstons, Rhodeses, Hobergs, and Howards—moved west not by covered wagon but by train and Model T. They were a restless, forward-looking bunch, and they traveled lightly. James Sr. and Agnes Inglis came over from Scotland in 1847 with five children and had six more as they made a slow westward progression through Connecticut, Ohio, Wisconsin, and Iowa. The Taylors, stolid New Englanders since the seventeenth century, suddenly uprooted themselves from the northernmost county in Maine and moved to Wisconsin in the 1880s; some of the younger folk then moved on to California. At age sixteen, my father's grandfather left the ancestral home of Hökhult (Hawk's Hold) in Sweden. He ran a restaurant in Seattle before heading to San Francisco shortly before the turn of the century. One couple got married in Iowa in the morning and left for California on the train in the afternoon; they arrived eleven days later, the honeymoon over. (It was only a five- to seven-day trip, so they must have stopped over somewhere, likely at one of the roadhouses maintained by the railroads.) The garage owner in Phoenix took his family to California in a Model T, crossing the Colorado River by ferry.

Like most of the people who arrived in L.A. in the first land boom, my forefathers were small business owners and tradesmen from small towns in the Midwest. At a time when most Americans were farmers, they were photographers, reporters, teachers, salesmen, grocers, bricklayers, shoemakers, coopers—people who could pick up and go, move with the times. The barrel maker didn't let his blind eye keep him out of the Civil War; instead, he hired a stand-in to take his physical and went off with the Ninth Michigan Infantry. The war correspondent accompanied General Sherman on his march to the sea and sent his reports back to New York via the new-fangled telegraph, which allowed on-the-spot war coverage for the first time. The owner of a meat market in Wisconsin became a local celebrity when he installed the first ice machine in town. After the war, the shoemaker—his profession made obsolete by the availability of factory-made shoes—moved out west, trained himself as a photographer, and opened a portrait studio. His son owned a bicycle shop in the 1890s; when he saw that the bicycle craze was waning and more people were acquiring automobiles, he turned his shop into Phoenix's first auto repair garage.

The men in my family were builders, whether they were developing a new business or a new community or actually laying brick and hammering house frames. The Swedes made a good living doing masonry contracting in San Francisco after the 1906 earthquake and conflagration, when business owners were eager to rebuild in fireproof brick. In his World War I–era newspapers, Franklin Johnston, with the blend of moral rectitude and boosterism typical of small-town newspaper publishers, extolled the virtues of clean, uncrowded beach living, just a trolley car ride away from downtown Los Angeles.

These men succeeded despite a lack of formal education; until my younger brother graduated from college, not one of the men in our family had ever attended a four-year college. My paternal great-grandfather Franklin Johnston left school at the age of ten to become a printer's apprentice. He educated himself by reading everything he could get his hands on, eventually building up an extensive library. He started his first newspaper at age sixteen and became a successful stockbroker, bank president, and newspaper publisher while still a young man. My maternal grandfather, Buford Rhodes, went to school through the eighth grade. He worked to pay for his brother's expenses at dental school and, later, for his fiancée's college education. Still, he managed to become a successful salesman with enough savvy to parlay his investment in several apartment houses into enough money for a more-than-comfortable retirement.

All the more unexpected then, is the fact that many of the women in our family were well educated. They had made sacrifices to continue their education. On my mother's side: Mattie (Mallicoat) Inglis left home in Greenfield, Missouri, in the 1870s to attend a finishing school at Osage Mission, Kansas. (The school, originally a Catholic mission for the Osage Indians, was recruiting white students to replace the Indian youths who had been removed to a reservation in Oklahoma Territory.) Mattie's daughter Grace later defied her father's wishes and entered Pomona College in the 1890s. Grace's daughter Dorothy graduated from UCLA in the 1920s with a degree in Romance languages, supporting her studies by tutoring other students; she later went on to do advanced study at the University of Mexico.

The women on my father's side followed roughly the same timeline. In the 1870s Lenora Taylor stayed behind in Maine when her family moved to Wisconsin so she could go to teacher's college; she taught for several years before following her family to the small town of Iola, where she opened a milliner's shop. Her well-to-do family was scandalized when she married a newspaperman. According to their daughter Smokey, the family finally accepted him when Lenora returned many years later from California, "sparkling with diamonds (how she loved those symbols of triumph!) and crowned with willow plumes." Smokey attended both USC on the West Coast and Columbia University on the East Coast in the 1920s. She worked for many years as a reporter and editor for her father's newspapers.

It did not surprise me to find that, for at least six generations, the women in our family had been gifted singers and pianists. (We were a family of frustrated musicians: my mother's plans for a career as a pianist were cut short by her early marriage; *her* mother had longed to become an opera singer, an idea her mother quashed because it was unseemly for a woman to appear onstage.) I was surprised, however, to discover an unusual pattern of marriage and childbearing. Many of these well-educated women married in their middle to late twenties; they invariably wed handsome younger men without much schooling; and they bore only one or two children. My grandmother once told me she remembered seeing her husband-to-be for the first time as "a beautiful boy coming down the hill with his blond hair blowing."

My parents were the first in several generations to break this pattern. My mother was only eighteen at the time she married; my father was two years older than her and had two years of college, while she had none; together they raised seven children.

I wondered why the men in these photos seem like clear-eyed, gentle dreamers while the women look hard-mouthed and steely-eyed. I found at least one explanation in the story of Grace Eulia (Inglis) Howard, my great-grandmother on my mother's side.

Grace's history begins with her father, James Inglis Jr. Even though his parents had arrived in the state nearly a decade earlier, I regard James Jr. as the true founding pioneer of our family because he and his brothers were the first to settle in the Los Angeles area. His parents, James Sr. and Agnes, had been on the move for almost their entire adult lives. They left Kirkintilloch, Scotland, in 1847 and took thirty years to cross the continent. Their last move, when they were in their sixties, was from Iowa to California; of their nine living children, only two stayed behind in the Midwest.

Of the two who stayed, one was the eldest son, James Jr. He and his wife and five children lived in Osage Mission, Kansas, where he made a good living as a carpenter. He built houses in the summer and made fine walnut furniture in the winter. "And so," his daughter Grace wrote with the sense of regret that marked her throughout her long life, "we lived in this small town contented and happy until Papa's family began urging him to come to California." The land boom had hit Southern California and carpenters were making five dollars a day—double the wage in Kansas. Like generations of emigrants after him, James was lured to California by the promise of high pay and steady year-round work.

The eldest of five children, Grace was fifteen at the time her family moved out West. As an old woman, she wrote a remembrance of her mother, Mattie Inglis. In it she tells movingly of her mother's death after giving birth attended only by a neighbor who did not "take care of her rightly." James Jr. went on foot to locate the doctor, who was on his horse-and-buggy circuit around the Florence area. He returned with the doctor in the evening, and Grace sat by her mother's side as the doctor examined her. She heard her mother murmur, "Too late, too late." Mattie died the next day. The family was unable to find a wet nurse for the baby, and he lived only twenty-three days. What Grace does not mention in her narrative is that it was left to her, at age fifteen, to raise her little brothers and sisters, ages nine, seven, four, and two.

Grace entered Pomona College in 1895. She was twenty-one, and after six years of taking care of children, cooking, and cleaning, she felt that it was her younger sister's turn to take over. She worked her way through almost two years of school before her father found out that his brother was helping her out with college expenses. Ashamed that his own brother was giving his daughter assistance he could not provide, he made Grace quit school.

Grace (Inglis) and Charles Howard, Minneapolis MN, ca. 1901. From the author's collection.

Back she went to cooking and keeping house for the younger children for another six years.

By the time her youngest brother was fourteen years old and able to take care of himself, Grace was almost twenty-seven. Lonely and determined not to become a spinster keeping house for her father until he died, she began writing letters to members of her church back in Kansas. One young man, Charles Howard, who had joined the church after the Inglis family moved, replied with interesting, newsy letters written in a fine hand. Within the year, she had taken a train back to Chicago to marry this man she had never met. I wonder if he had been honest about his age in the letters or if she was surprised when she met him. Charles was only eighteen years old—eight years younger than Grace. The marriage lasted longer than seems reasonable based on their scanty knowledge of each other and their conflicting personalities.

I learned what little I know of Charles from a handful of pages torn from his diary, which I found among Grace's papers. I had been told only that he was a terrible person who betrayed his wife and child. Reading these precious few pages, I saw my great-grandfather for the first time—not as a monster, but as an attentive husband, loving father, and literate individual.

Here was a man after my own heart. Charles was an amateur photographer and developed his own pictures. He wrote of books he read and plays they attended. A music lover, he wrote about going down to "the Edison agency to buy a fine phonograph with extra horn and six records." He read religious books in Hebrew. (He must have been self-educated, because he started working at the *Minneapolis Times* as a proofreader at the age of seventeen.) He slept on the parlor floor ("nearly froze") when his wife was pregnant, and wrote of his relief at going back to sleep with her a month after the baby was born. He sat with Grace when she was ill, brought her little gifts, and did the housework when she was recovering from childbirth. One night he worked at the *Times* until four in the morning, yet walked the floor with his "Little Sweetheart" for two hours before going to bed. He sounds like a lovely man, but he was also a "bounder." (My grandmother had to explain to me that this old-fashioned term meant he bounded from one woman's bed to another.)

A few years after the birth of their daughter, Grace left Charles and went back out West. Her younger brothers and sisters were now grown, with good steady jobs. They rented a house together and offered Grace two rooms in exchange for working as their housekeeper. Less than eight years after escaping from her duties at home, Grace was back to cooking and cleaning for her siblings.

At some point, long after her daughter, Dorothy, had grown up and left home, Grace was able to move into her own apartment. Her life revolved around her church, where she did both paid and volunteer work. I was sixteen when she died at the age of ninety-eight. I remembered her as an irascible old woman who was always scolding the Freeman girls for wearing pants. No matter what we were doing, she invariably asked us why we weren't in the kitchen helping our mother. With her antiquated view of a woman's role (her gift every Christmas was a new apron for each of the girls), she seemed like a relic of another century, utterly out of touch with the 1960s. I might have understood her bristly nature better if I had known of her life of disappointment and relentless hard work.

Years later, after Dorothy died her mother's diaries surfaced in a pile of miscellaneous belongings no one claimed. Over the course of several days, I read the entire set of small bound journals. They give a thorough accounting of Grace's good works, totting up the number of Ladies Aid Society meetings attended, homebound people visited, and items sewn for charity. I imagined my small determined great-grandmother arriving at the pearly gates with these diaries under her arm, neatly bound with string.

I found myself laughing whenever I ran across an entry that sounded like the woman I remembered. Sitting by her window one day in the mid-1960s, she wrote, "Every time I look out, someone going by—some girls and women with those awful tight pants on." But I also caught an occasional glimpse of a side of my great-grandmother I had never seen before. In one of the small volumes, November first, All Saints Day, is circled. Beneath the date is scrawled: "This day always brings sad thoughts for our mother died." Grace was eighty-five years old at the time, but she still remembered with an awful clarity what it was like to be a fifteen-year-old watching her mother die.

With my own keen sense of loss at my mother's death, I understood Grace's pain. I was both gladdened by the sympathy I could feel for my unapproachable great-grandmother and saddened by the thought that I, too, may never get over my mother's death, even if I live to be an old woman. That night I dreamed of Grace lying at the bottom of a clear river at the end of a red line that lay coiled on shore. I pulled gently on the thread and she floated toward me underwater, becoming younger and younger until, when I lifted her from the river, she was a baby, all tender and damp. I called for a towel to wrap her in and cradled her in my arms.

Learning something of Grace's life helped me to make a connection between the present generation and those that had come before. Nevertheless, the gap would always be there, perhaps because our family in the Southland lacked the continuity the Burnses enjoyed in the Snake River Canyon. For well over fifty years after the homestead was established, little changed for the Burns family. The life Dooley lived was much the same as the life his father had known and his grandfather before him. They still used kerosene lamps, trekked to the outhouse, and spent their Saturday nights dancing in the schoolhouse.

In L.A., the fast-changing pace of life left little to connect us with earlier generations. Even the environment has changed. Although we lived within two miles of where my grandmother had grown up, the area was barely recognizable to her when she returned to visit fifteen years after moving to Alaska. "Bean fields," she said as we drove her around Torrance, where many aerospace industries had mushroomed. "This was all nothing but bean fields."

I try to imagine the reaction of Grace Inglis or her father, James Jr., to seeing Florence now. Grace would undoubtedly see worse things there than women wearing tight pants. Florence is an unincorporated area of Los Angeles County best known as the flashpoint of the 1992 riots. The videotaped beating of a truck driver, Reginald Denny, took place at the intersection of Florence and Normandie Avenues. The area is studded with boarded-up buildings and trash-filled empty lots populated by drug users, prostitutes, and gangs. The section bordered by Florence Avenue to the north and Firestone Avenue to the south is even poorer than neighboring Watts and rougher than Compton. The year after the riots, the unemployment in the area was a staggering 21 percent, about the same rate as in Watts at the time of the 1968 riots. In some ways, Florence is probably as inhospitable a place as it was in the 1880s when the Inglises arrived, although the situation is improving. Financial incentives have lured grocery chains and banks to the community, and many Latinos and Asians have been successful at starting small businesses. Unemployment for Florence-Firestone has sunk to a (still abysmal) 12.7 percent.

It is no accident that these minorities (soon to become a majority) are crowded into the less desirable parts of the city. L.A.'s first African American community was located where the Civic Center is now. As the city expanded, they were pushed further and further south until the downtown community merged with the other longtime black settlement, Mud Town, later to be called Watts. African Americans were kept out of the beach cities by restrictive covenants that stayed in place for decades. Even after the covenants were removed, discrimination lingered. In the mid-1960s a petition was circulated in our neighborhood protesting a homeowner's decision to rent his mother-in-law apartment to an African American family. The internment of Japanese Americans in World War II vacated hundreds of acres of productive farmland in the South Bay.

It is no wonder, then, that African Americans, Mexican Americans, and Asian Americans are well-nigh absent from my family's story; they were literally removed from the picture. Just as in Kansas, where the forced exodus of the Osage Indians provided land and new opportunities for the Inglis family, in Southern California my ancestors benefited from governmental and social policies that discriminated against minorities and provided an underclass willing to work as cheap labor.

Running beneath my family's story of their rise in fortune in the West is a parallel story of the near-invisible people who supported their lifestyle.

Around the turn of the century, even a moderate-income household could afford "help"—Scandinavian or Irish in the Midwest, Mexican or Asian in the West. In the early days of her marriage in Minneapolis, Grace had a hired girl named Nellie Riley who helped with the housework and "went home nights." Later, when she was housekeeper of the Inglis siblings' household in Los Angeles, she had a Mexican woman come in every week to do the wash. My grandmother Rhodes said that her love of the Spanish language, which she studied and taught for many years, began as a child as she sat listening to the washerwoman's melodious speech.

In a photo of the Johnston family taken at Venice Beach, the family members, including a cousin who lived with them, are wearing bathing costumes. (The cousin was one of the many permanent houseguests Smokey described as "freeloaders" from her mother's side of the family.) A woman in street clothes, obviously still on duty, stands behind her young charge. A note on the back identifies her as Smokey's governess, Nellie Ingoldsby, who educated her until she entered the University of Southern California preparatory high school in 1912. The fact that the governess was included in the photo and named on the back indicates that she was treated with some respect. (Charles Howard, on the other hand, recorded the name of their "new girl" in his diary only to note the terms at which he had hired her.) More often, there is just a glimpse of an unnamed servant, caught by accident in the margins of the picture.

In the corner of some photos of my father there is an arm in a white starched blouse or a sturdy, stockinged leg: the maid who cared for my father when he lived with his grandparents. In only one photo is her face shown; she appears savagely unhappy. She would not have looked out of place in a mental institution or even a women's prison. I can't imagine my great-grandmother entrusting her grandson to this person. It makes me wonder what type of mistress she was. Smokey wrote of never being able to please her stern New England mother, and I suspect the maid found her similarly impossible to satisfy.

Occasionally there is more direct evidence of the life of these servants, such as the wedding photo of the Johnstons' houseboy and his newly arrived bride in traditional Japanese dress. At the bottom of the photo is an affectionate inscription to then-twelve-year-old Smokey. Haijiro Ito was a theological student while in their employ and later became a Methodist minister. I find it slightly embarrassing that my great-grandparents had a man of such obvious intelligence working as their houseboy, even though it was probably the only position open to him in the pre–World War I era. A cloisonné-on-brass vase he gave Smokey sits on a shelf in my living room

along with the rather bizarre note that he was later poisoned by Japanese
agents of the Black Dragon Society because he refused to spy against the
United States.

Smokey disliked people who couldn't or wouldn't do things for themselves, and she was annoyed by people who were condescending to others.
It is a fair guess that she was uncomfortable with the idea of having servants.
Certainly it was part of her whole upper-class upbringing that she cast off
as soon as she left her parents' house. She dropped out of USC because she
couldn't stand sorority life. She became a hard-drinking, ambulance-chasing
newspaperwoman and married four different men, all of whom were blue-
collar workers. Eventually she moved away just as far as she could—an
island in Alaska, where she delighted in thumbing her nose at her mother's
injunction to keep a clean house in case the minister came to call. She had
rejected not only servants but also the appliances and modern conveniences
that replaced them, hauling water, cooking on a wood stove, and carrying
out her own chamber pot (called in Alaska parlance a "honeybucket").

Smokey once noted in her journal that their cabin would have fit inside
the living room of her childhood home. I've seen both the cabin on Square
Island and the substantial living room of the house in Pasadena: she wasn't
exaggerating. The Craftsman-style house was built by the Johnstons at
a time when Pasadena was known as a winter resort for millionaires. I
would have been reluctant to move from the handsome house with its
stone porch and broad wooden beams. But I think it speaks well of Franklin
Johnston that he chose to leave stuffy Pasadena for the beach bungalows and
sand dunes of Hermosa Beach. The move was also a return to newspaper
publishing; after making his fortune as a stock broker, Franklin bought the
two South Bay newspapers and moved his family there in 1914. (I suspect,
had it been my great-grandmother's choice, they would have stayed in
Pasadena.)

When my father took me to see the house in Pasadena, he was somewhat
embarrassed that I insisted on knocking on the door to ask if we could peek
inside. At other houses that were once in our family, both in the South Bay
and San Francisco, I have not hesitated to ask the current owner if I could
look inside. The responses have ranged from a quick glimpse through a half-
open door to a complete tour of both floors. I can't do that in Florence.
It's not the kind of neighborhood where strangers go knocking on doors.
In any case, my ancestors' small frame houses and orchards surely have
vanished beneath acres of graffiti-covered apartment buildings.

Still, I wonder what it would be like to go back to the site of the Inglis
home, which is also the site of my grandmother's earliest memory: seeing

her grandfather, James Inglis Jr., lying in bed in agony, his hands encased in huge white bandages. When a bucket of roofing tar he had set on the stove to soften overheated and burst into flames, James Jr. had carried the bucket outside to save the house, causing burns to his hands that later proved fatal. (How catastrophes imprint themselves on young brains: Liz's first memory of her house afire; my father's of the Long Beach earthquake of 1933; his father's of watching the city burn after the 1906 San Francisco earthquake.)

I would have liked to walk the same land James Jr. had covered in his search for the doctor when Mattie was in labor, or to trace it—in mere minutes—by car. I imagine him, too poor to own a horse, trailing the doctor on foot across the Florence area, past fields and orchards, half-finished houses on undeveloped town lots. This unfamiliar land must have seemed endless under the flat November sky, as he walked from a farm where the doctor was last seen to the next, where, with luck, the doctor may be still—if he hurried.

Florence, no longer a coherent, functioning community, has reverted to a kind of no-man's-land, a dumping ground for L.A.'s unwanted. I have chosen Florence to serve, too, as my virtual junkyard. The Freemans have no other family resting place for miscellaneous junk. Unlike rural Idahoans, suburbanites don't have a weedy back forty in which to dump old cars and washing machines. Without basements or attics, there is nowhere for the detritus of generations to accumulate.

If there had been a family junkyard for the Freeman kids to explore, we would have played Bonnie and Clyde in the Model T that the Rhodeses drove across the Colorado Desert on a road made of wooden planks. We would sort through racks of lead type and practice laying them out right to left and backward—lightning fast—like Smokey and her father used to do. On the floor of the barn, we would come across Charles Howard's Victrola; with a couple of strong turns of the crank, we would listen to his favorite comedy record, "Uncle Josh in a Fifth Avenue Bus," and laugh at how unfunny it was. Our dinghy would be stowed in the rafters overhead like Dooley's *Virgin Sturgeon*. In one of the horse stalls we would find a trunkful of old clothes that come apart at the touch: Father's navy uniforms, Mother's taffeta party gowns, our old snow suits and rubber boots with the flip-over latches. Outside, the air would be heady with the scent of Uncle Herman's orange blossoms, great-grandma Ingrid Hoberg's bees buzzing around them. Abalone shells and broken glass negatives would crunch under our feet. And if we were feeling angry, we would grab a plate from the barrel of factory rejects Smokey kept in her backyard for just such occasions, and throw it, hard, against a brick wall.

In the wide open Florence of my imagination, I can wander for hours, and everything around me is a link to my family's past. In the corner of one of the outbuildings is a pile of Grace's canning jars, some still full of pickled peaches. My stepgrandfather Freeman's tools hang in the woodshop instead of being left at an interstate rest stop on one of our cross-country trips. The Johnston's house from Hermosa Beach still stands, ancient and weather-beaten, on a hill; there, in a staved-in drawer, I find the sugar cubes shaped like clubs, spades, hearts, and diamonds that my father used to sneak from the drawer and take back to bed with him in his tiny bedroom (once a maid's room) off the kitchen. In Florence, I can even walk through our house on Knob Hill with all the rooms intact and the staircase going the right way.

My brothers and sisters and I discover my father's boarding school standing in the weeds and we delight in burning it down. I toss on the fire my mother's piano music I would never be good enough to play, dozens of aprons Grandma Grace made for us, and those impossible-to-iron cotton dresses with rickrack around the hem. I save the set of china Smokey designed with a cherry blossom pattern, the ship's lantern from a wreck on the Palos Verdes rocks, and the kimono my father brought my mother from Japan.

I get to decide what to keep and what to burn, what I can live without and what I will always need.

SEVEN

Waiting for Coyote

The Nez Perce tell the story of five swallow sisters who built a dam that kept the salmon from migrating upriver. Coyote came down the river and didn't like what he saw, so he turned himself into a baby floating along in a cradleboard. When the board washed up against the dam, he cried piteously until the sisters found him. They fussed over the poor baby and took him home. But the next day, while they were out digging for camas roots, Coyote took a digging stick and started scratching at the dam, the dirt flying out between his hind legs. He worked at it for five days, until finally the swallows returned and caught him destroying their handiwork. The sisters darted around Coyote, beating him about the head. He plucked them out of the air one by one, then calmly finished destroying the dam. From that time on, the salmon were free to swim upriver to spawn and to provide food for the people and bears. And Coyote made the swallow sisters promise to build mud nests on the cliffs every spring to herald the salmon's return.

Asotin is little more than a gas station, a school, a diner, the county historical museum, a ball park. It's not too hard to imagine these simple buildings replaced by tepees. This flat spot at the mouth of Asotin Creek has always been a good place for people, with a year-round creek, plenty of shade, and lots of eels. Hesutiin, the place of the eels.

A few miles downriver from Asotin, Swallows Nest looms above the Snake like a flying buttress or the prow of an ocean liner. Visible for miles, it is only from up close that you can see the columns of basalt that form its distinctive shape. The cliff was once plastered with swallows' nests made of silt taken from the wide mud flats along the river. Lower Granite Dam transformed this section of river into a reservoir that extends a mile south of the town. The mud flats gave way to levees lined with riprap. Most of

Nez Perce women drying eels along the Snake River, ca. 1890. From the collection of Richard Storch.

the swallows now build their nests further upriver where the river still runs swift and there are miles of muddy banks and undisturbed cliffs.

Swallows Nest is one of three things I always look for when I drive through Asotin on my way to the Burns Ranch. The second is across the river on the Idaho side: a huge rock formation called the Whale, complete with symmetrical columns resembling baleen in a big blue's mouth. The third feature is Dooley's tugboat, moored just beyond the town park. I'm always cheered by the sight of the *Caroline*—red boat, fat black smokestacks, paddle wheel spokes just as homey as a white picket fence. In the summer, geraniums bloom in the lifeboat that sits on the deck.

I was sorry to have missed the moving of the *Caroline* from her old docking place at the state park, where she had languished untended for years. Liz remembers the party atmosphere that day, as friends came aboard the *Caroline* for the ride. A flotilla of jet boats pushed and pulled the old tugboat up the river, with kids running along the shore waving as they passed. Typically, Dooley remembers it differently

"Boy, was it ever a nightmare. You get that many people and they don't know what they're doing and they get into all kinds of trouble. We got into a hell of a mess. It was just crazy. Just like six people driving a car, you can imagine what happens. I couldn't control them. Somebody'd be pulling the wrong direction all the time, and then they'd get to zigzagging. I pushed barges for twenty years. I had considerable experience captainin' boats. But they didn't listen to me and I couldn't control them."

Even though the event had taken place years ago, I could hear the frustration in his voice. No wonder the move was a nightmarish experience for him: at the helm of a boat without power, unable to keep command of the crew, when he used to captain a tug boat with such precision that he could maneuver a barge over a predetermined drilling spot in the river and make it stay there. Holding the barge in place was a complicated process requiring four winches, ten anchors, and a knowledge of both the river current and local wind conditions. That was in the early 1960s, when geologists were investigating possible dam sites along the Snake and Clearwater Rivers.

If things had gone according to plan, there would be a fourth landmark at Asotin that would dwarf the other three: a dam as tall as a twenty-story building. Between 1962 and 1988 plans for a dam were put on the table repeatedly and then taken off again; dam proponents grew ever more frustrated each time as the full plate they imagined set in front of them was whisked away before they could take a bite.

To make sure there is a boat on our side of the river, we always stop in Asotin to call the Burns Ranch. The old-fashioned pay phone rings through before any coins are deposited. As I often do, I start speaking as soon as Liz answers, forgetting that she can't hear me until I insert a quarter. Ambrose skips into the market to buy a candy bar or a corn dog for the last leg of the trip.

If we kept driving past the ranch, eventually we would wind up at Heller Bar, where the road ends. To see the river before it disappears into Hells Canyon, we would have to drive up and over the Salmon River Mountains of central Idaho or the Blue Mountains of Oregon. Leaving its source high in the mountains of western Wyoming, the Snake flows over eleven dams before it reaches the Burns Ranch. On a map, the slash marks indicating dams appear at regular intervals throughout the length of the Snake, with one exception: the stretch between Hells Canyon Dam and Lower Granite Dam downriver from Lewiston. It is those precious 110 miles—the longest remaining free-flowing stretch of the Snake—that I care so much about.

The Snake River south of Weiser looks very different from the river I'm used to. In some places, it can disappear entirely. One hundred percent of the river is allocated to meet the demands for irrigation and city water. The crops grown in this arid region (especially sugar beets, which are more suited to the humid South) require large amounts of water. Add the growing demand for water from the booming population centers around Boise and Twin Falls, and it is no wonder that Idaho is the state with the largest per capita water use in the country.

In low-flow years, the river is sucked dry. Miles downriver it is reconstituted like cheap orange juice from treated wastewater, agricultural runoff, and a foul-smelling sludge flushed from the fish tanks at the local trout farms. The river used to be replenished by water that cascaded from the cliffs above the river at Thousand Springs; but the spring water is cold, clean, and clear—a perfect watery environment for trout—so fish farms have sprung up all along this stretch of river. Like the potato farmers, the trout farmers point out that most of the water they use winds up back in the river. What they don't mention is that the irrigation water is returned full of fertilizer, silt, and pesticides, and the water from the trout farms is released into the Snake full of extra nutrients that encourage the growth of algae. The surface of the slow-moving water behind Brownlee Dam is often covered with a green scum.

In recent years there has arisen a cry for more water in the river (mostly from downriver people who, the upriver folks believe, are always whining about something). Researchers are trying to determine what and how to feed the trout to reduce the amount of waste that ends up in the river. Farmers, unfortunately, are reluctant to experiment with water conservation because they are still operating under the outdated water rights system that says if they don't use it, they lose it.

People used to visit Shoshone Falls the way they visit Niagara: believing they are going to see one of the great wonders of this continent. Now visitors who find their way to the falls are often disappointed by the mere trickle that remains after all that water has been siphoned off upriver. But occasionally the falls returns to its former splendor, at least temporarily. I've stood at Shoshone Falls in the spring when there was a large snowpack in the mountains just starting to melt. The water thundered over the brim, throwing up rainbows. The mist rising in the warm sun made the trees look greener, the walls of the canyon redder, the people softer, more lovely.

When I get discouraged about the state of the river, I welcome Dooley's ever-optimistic outlook. He assures me that the water is cleaner than back in the old days when untreated sewage was released directly into the river. "Back then," he said, "it was a running sewer full of moss and bugs and trash fish: squaw fish, chiselmouths, red horses, shiners, the works. We used to feed them trash fish to the hogs. You think hogs smell bad," Dooley laughed, "well you oughta smell hogs that've been living on fish!"

These days, the Snake River doesn't really recover until it passes through Hells Canyon. Gathering speed as it falls in elevation, it aerates and cleanses itself as it tumbles over the rocks. Infusions of clean water enter from the Salmon River on the Idaho side and the Imnaha and Grande Ronde on the Oregon side. By the time it reaches the Burns Ranch, the river is clear, cold, and fast-running.

A few miles from the ranch, I slow down as the road turns to gravel. The canyon is empty. There are no boats on the river, no cars on the road. I love this time of year when even the swallows have gone. The sky is clouded over and the water is a still, silvery green. Spots of color are vivid against the muted background: yellow willows along the shore, brilliant green moss on the rocks, red sumac in the draws. A lone steelhead fisherman wades into the water, arms raised as if going to meet a dancing partner.

I know I will never share a fisherman's love of water. I'll never be a river runner like those who float down the Salmon and on into the Snake, taking their rubber rafts, dories, and kayaks out at Heller Bar, where the road ends (or begins, depending which direction you're headed). Never mind that I took swimming lessons every year of my childhood and that I learned to water-ski, handle a dinghy, and drive a ski boat. It makes little difference that I had surfer boyfriends with chapped lips and callused knees, a brother who jogged down the beach every morning with a squad of Junior Lifeguards, a father who learned to mend sails in the navy, and Yankee trader forefathers who sailed all the way to the Far East. No matter that we explored tide pools on the Palos Verdes Peninsula and camped on the beach in Baja; that we made sojourns to the freshwater lakes of Lake Shasta and Clear Lake; the man-made reservoirs of Lake Elsinore, Pyramid Lake, Lake Piru, Castaic Lake, Lake Cachuma, Nacimiento Reservoir, Lake Isabella, Pine Flat Reservoir; the saline lakes of Mono Lake, Salton Sea, Great Salt Lake; the mountain waters of the Kings, Kern, American, Truckee, Tuolomne Rivers; and the desert shores of the Colorado. Even though I spent most of my childhood near, on, or in water, I remain, stubbornly, a land person.

Although I have not been drawn to water, neither have I been afraid of it. My father made sure of that. He bought a used rowboat that he strapped to the car roof whenever we went camping. On the first of those trips, he took the five oldest kids out in the boat, leaving my mom standing on the shore with the baby in her arms and my little brother by her side. We rowed out to the middle of the lake. At my father's signal, we held on tight and starting rocked the boat, delirious with anticipation. With a wild grin, my father starting counting; when he reached three, he threw his weight to one side. We were cut off in mid-scream as the lake heaved up with a cold smack and we landed upside down in the dark green water. The life jackets popped us to the surface like orange balloons. We dog-paddled around the overturned dinghy while onshore several men raced for their boats until my mother waved them away. A small crowd gathered as we worked together to right the boat, haul each other in, and row to shore. The men laughed, "You had us scared for a minute there," and the women shook their heads and smiled at my mom as if she were the brave one.

It is all the more surprising, then, that I am often afraid of crossing the Snake. Even after years of rowing across to the ranch, the prospect of making that crossing lends a low level of anxiety to what would otherwise be a very peaceful drive. Even as I follow the curves of this familiar road, I'm thinking already about getting from point A to point B before the current carries me to point Z somewhere down around Lewiston.

When I first launch the Burnses' rowboat, my attention is taken up with bailing, maneuvering out into the river and getting oriented, setting a rhythm. But about midway I'm sometimes overtaken by a fear so intense it is almost paralyzing. My mouth goes dry. My arms are weak. The oars go every which way. Rushing water sounds suddenly behind my back, as if a whirlpool has just formed ahead of the boat. I glance nervously over my shoulder and it's just the creek tumbling down the rocky bank, telling me to steer slightly north so I will land on the beach.

One spring I took a jet boat trip up the Snake on a tour sponsored by the geology club at the local state college. The stops at the various geologic features were fascinating, but the boat ride itself was disturbing. Most travelers know the canyon from this perspective: sitting high above the water in mid-river, an engine roaring behind them. But to me, it was unexpectedly disorienting. This most earsplitting of river trips seemed strangely silent. All the small noises were missing: the whistle of a canyon wren, oars grating in the oarlocks, water against the rocks, the flip of a fish. I felt removed from my surroundings, as if I were watching it on TV with the sound turned down.

As we passed the Burns Ranch, I was surprised to see how tiny the rowboats appeared on the beach. I leaned forward in my seat to ask the captain if he knew the Burns family.

"Not really," he shouted above the motor, "but I've seen them out there rowing across in all kinds of weather. One time it was real high water, and I came up on the daughter rowing like crazy and not getting anywhere. I asked her if she wanted a tow."

I laughed, imagining Liz's reaction. He nodded. "I couldn't catch what she said, but man, I knew she was telling me to get lost."

Expecting him to pooh-pooh my fears, I admitted that I was sometimes afraid when rowing myself across at Burns Ferry. Surely, I thought, this captain with many years' experience had encountered plenty of scarier things than crossing the river in a rowboat.

"Well," he said, making that squirching sound out of the side of his mouth that farmers make when you ask them about the weather, "those boats are might-y small." And he told me about the time his fourteen-foot raft got sucked halfway down into an eddy before popping loose. "I thought I was a goner," he said, suddenly going quiet as he steered the jet boat through a set of rapids.

His comments made me think it all the more remarkable that Liz braves the river year round, coping with bad weather and extreme water levels. Liz told me how she had rescued the rowboats when they were broken loose in high water. It was early spring and chunks of ice were floating in the river along with branches and logs and entire trees swept down from the Salmon and Grande Ronde. The boats were her lifeline to the world; if she lost them, she would be stuck on the ranch for who knows how long because the road to Waha was an impassable mud rut.

Liz tried to retrieve one of the boats with her kayak but found the half-swamped boat too heavy to tow. She then jumped into her truck and followed the boat as it was swept downriver. When it came a little closer to shore, she leapt into the icy river and swam out to the boat. In water that cold, even a swimmer in good condition has only a fifty-fifty chance of being able to swim fifty feet (about two-thirds the length of a basketball court) without succumbing to the cold. Liz managed to reach the boat, climb in, and row to the Washington side. She waited at the garage for Bertie to get back from town. By the time Bertie arrived, Liz was blue with cold, her teeth chattering so hard she could barely talk. Bertie slipped out of her coveralls and boots and helped Liz put on the dry clothes. Bertie bailed the boat and rowed them across, battling a current so strong that they were swept all the way down to the dump before they could beach the boat. It was a long cold walk back to the house.

I've never attempted to row myself across the Snake at spring flood, but I've sat very still in the back of the boat as logs and chunks of ice barreled past, a kaleidoscope of whirlpools forming and unforming around us. Equally as dangerous as the whirlpools are the boils that well up suddenly from beneath, spitting out huge chunks of debris that can knock an oar out of your hand, batter your boat, and spin you around in the rushing water, disoriented. Ram's Head Rock was notorious for its dangerous eddies, yet the Burns family risked the crossing every day when they lived at the north end of the ranch. For Dooley and Liz, rowing across at Burns Ferry is a piece of cake compared to Ram's Head. To me, it is excitement enough.

In high water a large eddy forms at the wide bend in the river at Burns Ferry. The eddy extends from shore to shore and rotates counterclockwise so that in front of the boat launch on the Washington side the direction of the water flow is reversed and the water moves at a fast clip back upriver. One day in early spring we stood on the road looking down at the whirlpool; it was like gazing into the center of a hurricane, complete with a windlike roaring. The children were excited and scared, and I tightened the straps on their life jackets until they complained. On the Idaho side, Liz moved out into the river, headed slightly upriver; cutting back the power, she let the boat be pulled into the outer edges of the eddy. The boat moved swiftly in a giant arc toward the opposite side. Liz used the momentum of the eddy to sling the boat toward the shore, then she turned the boat around and powered against the reversed current and landed neatly at our feet. My husband said, "That was very Zen." She laughed and said, "You gotta go with it, not against it. You fight it, it gets you every time."

The next day Bertie took us back across the river. The water had dropped a little, but the eddy was still circling shore to shore. When the boat was caught by the current, she opened the engine full throttle and powered directly across the whirlpool. We hung on tight as the boat dropped into the vortex, holding our breath until it climbed back out again. Bertie had become adept at calculating the strength of the eddy, but her frontal assault lacked the grace of our crossing the day before. Liz had caught that current and let it fling her like the grandmothers were playing crack the whip; to stay on her feet, she had to know just when to let go.

On the way to town, there are many stretches of road built on fill dumped at the foot of the cliffs by early miners anxious to create an alternate route to town that did not involve climbing the precipitous trails along the ridge tops. When the river is especially high it often covers these low spots, the current running fast against the steep bluffs. Liz or Dooley trust their ability to read the water, but Bertie in her little Toyota requires more solid assurance that the gravel road hasn't washed away. She admits, "I'm chicken. I wait for a big pickup to come along and see if it makes it through." I don't think I would risk it, knowing that most people who perish in floods die in precisely such a manner: trying to drive across a flooded road after misjudging how deep and fast the water is. But even these high flows don't match the volume of water that used to come through here in the spring. Now much of the water is held back to fill the three upriver reservoirs for the summer months. Before the dams, spring snowmelt charged down the canyon at between fifty thousand and seventy thousand cubic feet per second, almost five times as much as the current flow.

The thought of anyone attempting to cross the river in that kind of high water is almost inconceivable. And yet, less than thirty miles upriver is the point, marked now by a small sign, where Chief Joseph's band of Nez Perce was forced to cross the Snake at spring flood. In the spring months, the runoff from the high country of the Wallowas and the Seven Devils is cold enough to stop a person's heart. Chief Joseph's band, from what is now northeastern Oregon, was given just a few weeks to round up their cattle, pack their possessions, get across the Snake River, and make their way to the reservation at Lapwai, Idaho. Homesteaders who had steadily encroached upon their homeland in the stunningly beautiful Wallowa Valley wanted the Indians out. Nez Perce families loaded their possessions into bull boats made of hide and herded over six thousand head of cattle and horses into the swift water. Hundreds of men, women, old people, and children crossed the river on horseback or in the round boats. Some of the stock drowned, but incredibly, no human lives were lost.

Yesterday Liz called to tell me that Dooley's friend Lon had just come in to say they got an elk, and he was taking the horses back up the canyon to pack it out. It was just after dark and Liz was concerned about Pop, who had stayed to skin the elk. I imagined Dooley alone on the mountain, hunkered down near the still-steaming body of the huge animal. Dooley's walk was not as steady as it used to be, and his hands shook so he had trouble doing simple tasks like installing the spark plugs on his tractor engine. I wondered how he could aim well enough to hunt. "You'd be surprised," Liz said. "Pop can't go up the mountain after elk like he used to, but if he's got someone along who can move one in his direction, he can shoot it dead on every time." Ambrose and I were invited to the ranch to help cut up meat for the freezer, clean up a summer's worth of accumulated detritus around the barn for a bonfire, and generally batten things down for another winter.

In the schoolhouse, a pair of upturned antelope hooves form a coat rack. Above the rack, "Burns Ferry Schoolhouse, District 13" is penciled on the plank wall. In the main room, Dooley sharpens several large knives, takes the meat grinder apart, and washes the parts in soapy water. A pile of potatoes have been dumped in a corner, still covered with dirt and smelling like stones. The walls, lined floor to ceiling with shelves, are like the pages of a children's book that asks the reader to find certain objects: a grinding stone, a piece of driftwood shaped like a hissing cat, a dead bat. There are things here that must have belonged to Liz's mother: large flower vases, a teapot, a spice rack, a box of sewing notions, including lengths of ugly rickrack that, Liz and I have agreed, all mothers in the mid-1960s felt compelled to apply to every item of girls' clothing. From a nail dangles a fishhook the size of a man's hand, the six-inch-long shank twisted by the thrashing of a powerful sturgeon.

There are not as many books as I remember from previous years. A lot of them have been chewed up by pack rats. Dooley claims people have walked off with many of his books, but Liz remembers him stuffing armfuls of paperbacks into the walls for insulation. Still, there are hundreds of books arranged in rough order by subject: large volumes on Japanese art, Northwest geology, birds, plants, ghost towns. On the other side are books of fiction, an assortment of books on science and history, including a volume of speeches by Nobel Prize winners, and a stack of *Scientific Americans*. (Dooley and my father are the only people I've ever known who read—and understood—*Scientific American*.)

On the west wall of the schoolroom are Liz's old books: *Pinocchio*, *Heidi*, *The Bobbsey Twins*, a whole shelf of Nancy Drew. Below the window are books from Dooley's childhood, including a set called the Young Folks Library, published in 1902: *The Book of Great Sea Stories*, *The Book of Famous Explorers*, *The Book of World Travelers*, *The Book of Brave Deeds*. I wonder if the stories in these books helped sustain a girl who often had to face dangers both real and imagined. I know that one of young Lizzie's favorites was Rudyard Kipling's *Captains Courageous*, which tells the story of a spoiled brat who is pressed into service on a sailing ship and learns to tough it out.

Liz told Ambrose once, "Courage is like a muscle. You have to use it regularly on little stuff, so you'll be ready for the big stuff when it comes along." As a child she was sent high on the breaks after cattle, riding as much as twelve hours before nightfall. Like many of the canyon residents, the Burnses never carried water, believing that drinking a little bit just makes you more thirsty. Young Liz was told, "If you get to wantin' a drink, suck on a pebble." She climbed heights to locate lost cattle and herded the half-wild beasts down the mountain. As dusk fell, she may have been still finding her way back from an unfamiliar creek drainage miles from home.

Liz had to contend with many fears when she was on the mountain. At home, she had just one, or rather, a flock of them: chickens. From the time she was very young, it had been her job to collect the eggs. If she was afraid of how the chickens flew at her and pecked her hands, well, her parents said, she had just better get used to it.

As a child, there was just one thing that scared me more than the Zodiac Killer who roamed the Southland: being on my own. In the ranching culture, resourcefulness and independence are qualities that adults value in children. They may be leading no one but a couple of stubborn cows, but ranch kids are leaders all the same. In suburbia, kids are raised to be joiners and followers. We formed ourselves into cliques, joined clubs, surrounded ourselves with others like us. We learned to follow directions, rules, each other. At home, my siblings and I were not encouraged to think for ourselves or to take the initiative. The Freeman kids learned to rely on each other and to find strength in our numbers. We functioned best as a unit. Without my six siblings by my side, I felt puny and alone, a mere splinter of the mighty oak that was our family. Making the crossing to the Burns Ranch is frightening because when I'm on the river, I'm on my own. Even with Ambrose in the bow, I'm alone at the oars.

What remains of the elk is lying on a table in the main schoolroom. I trim the fat off a chunk of meat the size of a toaster oven, Ambrose cuts the chunks into strips, and Dooley feeds them into the meat grinder, poking it down with a wooden spoon. We just took some ground elk into the house for Liz's approval. She likes it lean and Dooley likes it fat and juicy, so they had to reach a compromise. She fried up a small patty and we tossed crispy bits of it from hand to hand before popping in our mouths. It was hot, rich, a bit greasy, just perfect.

It is cold and dim in the schoolhouse. I keep picking pine needles and elk hair off the meat. It's probably a good thing there isn't much light coming from either the grimy windows or the bare bulb dangling over the table. Liz had warned me, "You never know what all ends up in there when Pop does the grinding." But that doesn't stop her from eating it. The meat has never made her sick, but neither is she affected by a lot of things on the ranch that would make other people ill. She eats fruit off the ground without washing it, drinks water from seeps, plucks ticks off herself regularly. She sweeps out the barn, breathing dust that may be infected with the Hanta virus, and lets bats fly in and out of the house without worrying about rabies. She walks around barefoot over ground that's a minefield of thistles, burrs, rusted nails, rocks, bone shards, thorny sticks, and ever-present reminders of dogs, horses, cows, pigs, and turkeys. She doesn't step gingerly, saying eech, ouch, ooch, like I would, but walks deliberately flat-footed like an Indian.

I commented on her iron constitution when she stopped on a hot day to drink from a broken water pipe. "Yup," she said, wiping what appeared to be equal parts mud and rust from around her mouth, "I have a scientist friend who says she can't believe all the microorganisms I ingest and still survive." Scientific studies have suggested that the rise of childhood asthma in recent years is due to Americans' obsession with cleanliness. It turns out that children like Liz who are exposed to plenty of dirt, animals, and germs rarely develop asthma or allergies.

The Nez Perce children raised on these lands developed a similar resistance, but they were vulnerable to the germs brought by the whites. Dooley mentions that children of the Nez Perce had chronically runny noses. He jerks his thumb at Ambrose. "When I was about the size a him, we always played with the Indian kids." The handful of kids at isolated Burns Ferry eagerly awaited the arrival of the Nez Perce children each spring. A band from the reservation traveled on horseback along Lapwai Creek to the top of Craig Mountain and then continued down Thornbush Creek to their traditional fishing grounds on the Snake River. It was a thrilling moment when they were spotted coming down the trail, the women riding with a toddler in front and a baby on their back, the men driving the winter-starved ponies, and the children running alongside.

The band, including anywhere from fifty to seventy-five people, camped at Thornbush Creek for several weeks to catch sucker fish, salmon, and eel. After about a month, the band brought their horses down from the high pastures, stored their tent poles in the caves along the creek, and headed out. Some went back to the Salmon River and others up to Chief Joseph's country in the Wallowas, as Dooley said, "like they had since the beginning of time."

Eventually Dooley's friends grew up and returned with their kids, who became Liz's playmates. Like the Nez Perce children, Liz learned to swim before she could walk, paddling around with the others playing on the gravel bars while their mothers washed clothes in the river. She remembers sleeping with them in the teepees with the edges rolled up to catch the night breeze.

"Why did they stop coming?"

"No fish," Dooley said. "Got Brownlee Dam built up there and it stopped the chinook runs long before these other dams were thought about."

Brownlee went into operation in 1959. Within two years, the number of salmon caught in a night of dip-netting went from fifteen or twenty to maybe one, sometimes none.

Between Dooley's workshop and the beach, there are waist-high rock walls that have been there for centuries. They look like terraced garden walls angling toward the river. The ancestors of the Nez Perce used to anchor their boats over the walls and net the fish as they came upriver. Over the years, as silt from the creek built up the bank, the river has gradually moved west, leaving the rock structures high and dry. The Indians would then build a new wall, parallel to the old one. Now the river covers the lowest, newest wall only in the spring when the water is very high. If the salmon still ran in great numbers and the Nez Perce returned to their traditional fishing site near the mouth of the creek, it might be about time to build a new fish wall.

"In the dark at night," Dooley said, "the fish come along right in close to shore. The fish would come up and follow along the wall, and you'd dip down like this and feel them hit the net, and you'd jerk it up. Clear on up through the fifties, why, we was a-catchin' all of them we wanted."

Liz remembers curling up in the bottom of the boat, rocked to sleep by the motion of the water. She awoke crying each time a salmon was hauled in and the huge wet tail slapped her in the face. At three or four years old, she was just about the same size as the chinook.

It was known at the time they were built that the dams would have a negative impact on the sizable commercial salmon industry, but people were dazzled by visions of the underpopulated, underdeveloped Northwest suddenly transformed: skilled workers would flock to the modern new industries in the cities, and farm families would grab up the irrigated land, eager for the challenge of making the desert bloom.

There are a total of eight dams between here and the sea: Bonneville, The Dalles, John Day, McNary, Ice Harbor, Lower Monumental, Little Goose, and Lower Granite. My father's stepfather (the man who gave us the Freeman name) worked on the construction of Bonneville Dam in the 1930s. Like many of the men in our family, he was a builder, and he was proud to work on a project that would bring progress to his native Northwest. A heavy equipment operator, he later went on to level the sand hills above the South Bay for home sites and bulldoze roads through the jungle in the South Pacific as a Seabee in World War II. Bonneville, the first dam in the Snake-Columbia River complex of dams, has been of unquestioned use in providing power and protecting Portland from the devastating floods it had experienced in the past. The need for Lower Granite Dam is not as clear.

Pullman is the town closest to Lower Granite, the last dam built on the Lower Snake, roughly a half-hour drive from our house. The road down from the Palouse's high plateau reaches the river at Almota, where grain is loaded onto barges. In bumper crop years, the elevators fill up and wheat is simply poured onto the ground like golden sand from an hourglass. Just upriver from Almota is Boyer Park, one of the county and state parks strung every fifty miles or so along the Lower Snake. Like its even lesser used counterparts, Wawawai, Glens Ferry, and Central Ferry (which is reached via the aptly named town of Dusty), Boyer Park is an incongruous patch of green among the brown hills. Thousands of gallons of water are pumped from the river each year to provide shade and a grassy lawn for the campers in RVs. It costs so much to maintain these islands of green that the state keeps threatening to close the parks.

Just below Boyer Park is Lower Granite Dam. When we drive over the dam to reach the dunes, we often stop to watch the locks fill with water or to watch steelhead swim by from the viewing room. The dam supplies little electricity and its turbines run just a few hours every day. Its primary function is to boost grain barges the last hundred feet needed to bring them up to the level of Lewiston. By the 1970s salmon were already being pressured by overfishing and habitat degradation, and each dam built on the Snake had made their journey upriver that much more difficult. But Lower Granite is widely believed to be the straw that broke the salmon's back. Environmentalists had tried to stop the dam, pointing out that its electricity wasn't needed, but advocates of the project pushed for completion so that Lewiston, 450 river miles from the Pacific Ocean, could become a seaport.

In an attempt to restore salmon to historic levels, federal and state agencies as well as tribal governments have already tried everything short of dam breaching, including a controversial "drawdown" of the river in 1992. To mimic the effect of a free-flowing river, water levels were lowered behind Lower Granite and Little Goose Dams; the reservoirs became a river again, speeding the smolts toward the ocean.

For the first time in almost twenty years, the Snake ran swiftly in a narrow channel. Long-buried structures and landforms began to reappear. Rotting pilings. A rock formation called Cabbage Head. Nez Perce landmarks: a rock called Coldweather Girl, where Winter's Daughter stood in the water; a waterfall formed when Coyote extended his penis across the river to impress some girls. Mudflats. Cars. Dead fish. Gravel bars. Tires. Fishing gear. Forgotten swimming holes. An island once connected to Lewiston by a footbridge, where families would sit on bleachers to watch baseball games. An office safe. Handguns. Human bones. Another island, one that housed a tent city of hoboes in the 1930s. A calm back eddy where children used to ice-skate in winter. Docks cracking from their own weight. Swallows dipping up mud to plaster nests on the basalt cliffs. A river playing music again.

Unfortunately, without water to support structures such as docks, millions of dollars of damage was done. Those dependent on the reservoir system—farmers who relied on barges to get their grain to market, the pulp mill, the boating industry—called the drawdown a colossal failure. Even supporters of the plan were daunted by the cost of repairs. What was supposed to have been the beginning of a series of drawdowns became a one-time experiment.

The fish biologist at the Idaho Department of Fish and Game office in Lewiston briefed me on the salmon recovery efforts that had recently made possible a limited season on chinook salmon for the first time in twenty years. He was just another bureaucrat spouting statistics until I asked him what it had been like to participate in that rare season on the Clearwater River. His eyes lit up as he struggled to describe the feel of that chinook— what Dooley calls a "high-spirited fish"—on the end of his line. "It was like nothing I've ever experienced before."

I asked, "Do you think there's any hope for the salmon?"

He said seriously, but with that light still in his eyes, "Yes, I do. I have to believe that. Otherwise I wouldn't be able to come to work every day."

There are three dams on the Middle Snake: Brownlee, Oxbow, and Hells Canyon. Commuter flights between Lewiston and Boise routinely fly over Hells Canyon, and the dams are easy to spot: Brownlee, one of the largest rock-fill dams in the world, looks huge even from twenty-three thousand feet; just a short distance downriver is Oxbow Dam, placed at an omega-shaped bend in the river; Hells Canyon, wedged between cliffs about six hundred feet apart, appears as a tiny cinch pulling the canyon tight. In the reservoirs behind each dam, the Snake lies bloated and sluggish, covering the canyon floor. Below Hells Canyon Dam, the river shrinks to a bright ribbon winding back and forth across the bottom of the canyon, with beaches, islands, and gravel bars.

I try to imagine the river as it looked before the dams, when there were four times as many beaches in Hells Canyon. A longtime resident of Asotin told me, "What everyone used to remark on about the Snake was all the white sand beaches. Now they're all gone." The silt needed to replenish the beaches gets trapped behind the dams, and the daily, even hourly, fluctuation of the water level eats away at the riverbank.

Like the eight dams on the Lower Snake and Columbia, these three on the Middle Snake have had a catastrophic effect on the salmon runs. Salmon are the megafauna of the Northwest. They appear as iconic figures everywhere from art galleries to tackle shops. Colorful sockeyes and muscular chinook appear—tails flipped—on hats, T-shirts, billboards, and the covers of sportfishing magazines. You can find salmon-shaped pencil holders, flower vases, hat racks, wall sconces, serving platters. Salmon restoration efforts have cost hundreds of millions of dollars and have caused more debate than any other issue in the Northwest. With all the focus on salmon, the decline of other species has often been overlooked. Eels, too, once swarmed up the river every spring, and sturgeon cruised hundreds of miles out to sea and back again.

Sturgeon are large primitive-looking beasts that have survived virtually unchanged since the time of the dinosaurs. Over a lifetime that can span a hundred years, a sturgeon might travel dozens of times up and down the Columbia River system and out into the Pacific, going as far north as the coast of Alaska. At the turn of the century it was not uncommon for a commercial fisherman to haul in his set line and find a twelve-foot-long sturgeon weighing over seven hundred pounds. Most of the sturgeon caught these days are under five feet in length. Fish experts speculate that without the spawned-out salmon, there aren't enough nutrients left in the river. It's hard to imagine a sturgeon growing to record size by eating crawfish, which is now their primary food source. Or perhaps the sturgeon eventually get too big to negotiate the fish ladders on the lower Snake and Columbia River dams and remain in the ocean.

Now it is illegal to remove sturgeon from the water; fishermen will hold one alongside the boat, admiring its retractable shovel snout and the razor sharp ridges on its back, until it catches its breath. They let it go hoping that someday their children, or even their great-grandchildren, will catch that same fish, grown to the size of a canoe.

Everywhere you go along the Snake River there are pictures of these huge fish, like the one of a girl sitting on a saddle atop a sturgeon as big as a pony. There are plenty of stories, too, such as the time a homesteader caught a sturgeon so heavy that he hitched up his horse and boy to haul the thing out of the water and almost lost both stock and child when the monster made a run for the deep. As teenagers, Dooley and his brother earned money by catching good-size sturgeons—eight or nine feet long—to feed the sheepshearing crews. One especially large fish used to hang around the eddy at the mouth of Thornbush Creek. Dooley knew its exact length—fourteen feet, six inches—because he landed it once and measured it before releasing it again.

No one sees sturgeon that size anymore; they have moved into the realm

Sturgeon taken near Lewiston, 1930s. Nez Perce County Historical Society, Lewiston, Idaho.
74-36-1-11.

of our imagination and spawned their own tall tales. Nez Perce tell of "sociable" sturgeons that swam close to children in the water or, like a friendly dolphin, offered a fin to a drowning person and escorted him to the surface. Old-timers swear they knew the miner who claimed to have harnessed a sturgeon to two railroad ties and ridden it down the Snake River. They say he stood on the ties with a rope wrapped round one fist like a bronco rider, and rode that thing "from Brownlee to Copperfield."

I had never heard of sturgeon (except as the source of Russian caviar) before I moved to Idaho. But even as a child, I was familiar with moray eels. Beach residents were not big readers, and compared to other homes in the neighborhood, our house had a lot of books. (Looking back, I realize it was but a small fraction of the number of books in Dooley's schoolhouse.) I had an insatiable appetite for books, and I often ran out of reading material. The long bookshelf in our living room held a few nursery rhyme collections and my father's Book-of-the-Month Club hardbacks, which looked incredibly boring, and probably were, given the ease with which he used to fall asleep reading them.

I resorted to curling up with a volume of the *Encyclopedia Americana*, reading about foreign countries and surgical procedures. I found the single most disturbing picture in the entire encyclopedia—even worse than photos of rare skin diseases—was a photo of a moray eel's mouth lined with a circlet of jagged teeth. A moray eel was a nightmare creature, a leech with the teeth of a piranha.

My father and brothers used to go deep sea fishing from a barge out in Santa Monica Bay. They would bring home a barracuda with a jutting jawful of teeth, a halibut with both eyes on one side of its head, or a red snapper with a bladder of air protruding from its mouth. Occasionally, a fish would have a circular scar on its side: the mark of a moray eel.

Eels returned from their sojourns in the Pacific to fast-moving creeks all along the West Coast. They made their way upstream by attaching themselves by the mouth first to one boulder and then another. Nez Perce boys harvested eels in Asotin Creek by pulling them off the rocks, being careful not to grab them low on the neck. An eel with too much wiggle room could whip around and fasten itself to the boy's arm, leaving an ugly circular scar.

The fact that eels were a staple of the Nez Perce's traditional diet did not matter to the Army Corps of Engineers. They considered the eels merely pests and installed iron bars to keep them out of the fish ladders around the dams. Now the eels have stopped coming.

I ask Dooley about the morays, which locals call blue eels. Dooley holds his hands about three feet apart, "The big blues was about like that. I suppose they weighed a couple of pounds or more apiece. The Indians put them in a gunny sack, and it'd be so heavy they couldn't hardly lift it. If you put rocks in the sack and wet it, the eels would suck on the rocks and live until they got ready to clean them. They'd scrape and scrape and scrape." Dooley used the bloody knife in his hand to make a motion like stropping a razor, and I could see the Nez Perce women with their scrapers, sitting with their legs to the side, shawls round their shoulder, scarves covering their hair. "Get every speck of blood and everything off of them. They'd cut the eels twice and put little sticks inside to spread it wide. Put them up in the smokehouse they'd built in the thorn bushes and lay a blanket over it. They'd build a fire under them out of willow. Smoke them until they were just plumb dry."

Dooley hitches up his pants with his wrist and continues, "If they wanted to eat the eels right away, they'd get a green stick about as big as your finger and curl them up on the stick. They'd hold it over a small fire and grease just ran right out of those eels. They'd sit there and turn them till they got cooked. Didn't skin them or dress them or nothin'. Eat them just like people do sardines." He laughs and feeds a strip of meat into the grinder. It goes in with a horrible slurping sound.

The light is fading and our hands are so cold that Ambrose and I have trouble closing the Ziplock bags full of ground meat.

"They liked them big yellow-bellied suckers, too. They were just crazy about them things. Bony things, but, oh, they's so damn good. Sometimes they baked the fish. They'd dig a hole in the sand, build a fire in there and get it goin' real good. They'd put a whole bunch of suckers in, and immediately bury it with sand. And in an hour or more they'd go dig it out and take sticks and eat the thing. Never dressed the fish. They were crazy about the eggs in them. Those big old yellow-bellied suckers had a lot of great big long yellow skeins of eggs. That was the first thing they'd eat."

I think of all the meat stacked in the Burnses' freezer, enough to last through the winter and more. In the old days, after a long winter of dried fish, roots, berries, and moss soup, the Nez Perce must have welcomed the arrival of the suckers—the first fish to come up the river in early spring.

I would have liked to have been there on the beach when a suckerfish was dug from the hot sand. I would watch how the skull fell apart into separate pieces and listen to the stories of how the tiny, intricately shaped bones came to be called grizzly's earring, raven's socks, and cricket packing her child.

We finish with the elk. Over sixty pounds of meat. The dogs are lying outside, so satiated they ignore the hip joint I throw to them. We wash our hands and fill a bag with pears from a box in the corner.

The fish were not the only things affected by the dams. The character of the river has been changed in many large and small ways. The Snake no longer gets cold enough to freeze because the water released from the depths of the reservoir behind Hells Canyon Dam stays at a relatively constant temperature throughout the year.

The first time I heard about ice on the river, I imagined it like a scene from Moscow writer Carol Ryrie Brink's children's books about the early West: a young couple tucked beneath a buffalo robe taking a cozy ride across smooth ice, bells ringing on the horses' harnesses. But it wasn't like that at all. The ice broke up and then refroze, forming treacherous pressure ridges. Unable to get their supplies by boat, ranch families drove their horses across, pulling sleds loaded with groceries and other necessities. The horses often panicked on the slippery surface and plunged through the ice, cutting their legs to ribbons.

Old-timers say, "It doesn't get cold like that anymore," remembering the winter of 1948, when the paddle wheeler *Minerva* got stuck in the ice at Ram's Head Rock, and hay had to be flown in and airdropped to feed the stock. There is a sadness in their voices, not because they miss those days of hardship and cold, but because the world is changing in unfathomable ways. The weather is doing strange things. Farmers, who used to be heroes for feeding the world, are now being criticized for overproduction and blamed for causing erosion and water pollution. The need to control predators was once unquestioned, but now the national forests are being handed back to the wolves and the bears. The very idea of progress is being turned on its head. What many had thought unthinkable—breaching the dams—was being debated at town meetings and in letters to the editor in newspapers throughout the region.

Many area residents remember how the feds had come and talked them into building the dams in the first place, back in the 1960s. Yes, orchards would be bulldozed, houses moved, Nez Perce burial sites flooded, but it would all be worth it. Sure it wouldn't be so good for the salmon, but think of sportfishing in the reservoir: smallmouth bass! crappie! (Both non-native warm water species.) The dams would bring progress and prosperity to the region. Jobs would be generated just as surely as hydroelectricity from the turbines. Now the federal government is not so sure, telling locals a free-flowing river may be better after all.

To protect endangered runs of salmon and to honor fishing rights guaranteed by treaty, the dams might have to go. At a local town meeting about salmon restoration, a Palouse farmer stood up and shouted, "They talk about a free-flowing river! Doesn't anyone remember what the river was like before the dams? It was dangerous, there was no access, bust the bottom of your boat on it!" His words echoed those of another man who had just declared that if it weren't for the farmers, the land would be "just like it was when the Indians were here. They wouldn't have done a damn thing with it." I stopped following the debate, lost in the thought of the Snake River and the Palouse hills before a damn thing had been done to them.

Mary Jim, one of the last members of the Palouse tribe to live along the Lower Snake, remembered what it was like before the dams. The river once "played music to my people," she said in a newspaper interview many years ago, "but now the river is silent." What would it be like to have that sound silenced? I try to imagine my childhood nights without the fall of waves on the beach. I can't imagine it. The sound swung me in my sleep like a hammock.

The sound of the ocean surfaced after the TV was turned off and my mother stopped clattering the dinner dishes. It rose to fill the silence left after my father had ended his day with *Night on Bald Mountain* or the *1812 Overture*, turned up so loud it shook the floor. I heard it most clearly in the gray hour just before sunrise, but it was there in the background throughout the day, ever present as the ocean breeze that moved the curtain in our bathroom window. I knew that the sound of waves had been there through my mother's childhood, my father's childhood, and my grandmother's childhood. And it had been there throughout my great-grandfather's decade as an invalid with heart disease, reclining on a wicker chaise longue on the porch, years passing with little change except the face of the nurse who bent to tuck a blanket around his legs.

In 1958 the Army Corps of Engineers recommended that a dam be built at Asotin. The dam wasn't needed for power generation, but it would make it possible for barges to reach a major limestone deposit on the Idaho side. With all these dams being built on the Columbia River system, the engineers figured with a kind of circular logic, they were going to need that limestone to make cement. Congress gave the go-ahead for the Asotin Dam in 1962. The twenty-six-mile reservoir would back up slack water all the way from Asotin to the confluence of the Snake and Salmon Rivers. The river level would rise fifty feet, covering every building on the Burns Ranch as well as the garden, the winter pastures, the barns and the corrals.

Archaeologists in the Northwest had been doing frantic salvage digs for years, keeping just one step ahead of the dam builders. Many sites, including the famous Marmes Shelter in eastern Washington, were inundated before they could be thoroughly excavated. In 1964 a team of archaeologists set up camp at the Burns Ranch while they worked to recover artifacts from the area to be flooded by the Asotin Dam reservoir. They catalogued almost a hundred seasonal camp sites, twenty-seven house pit sites, forty-four burial sites, twenty-three storage shelters, and other evidence of human occupation, including fish walls, petroglyphs, and sweat lodges. The dam-building binge in the Northwest had been going on for thirty years and appeared to be unstoppable; there was little chance that even a significant archaeological site like petroglyph-covered Ram's Head Rock would be spared. The archaeologists knew they were just spitting in the wind, and their sense of futility is apparent in the report they filed: "During the survey we recorded a limited number of historic sites. Many more exist in and adjacent to the Asotin Reservoir area. These include mines, quarries, lime kilns, school houses, ranches, sawmills, and pioneer graves. We believe that some provision should be made for locating, describing, and photographing these sites of historic interest before they are destroyed" (*Archaeological Survey and Test, Asotin Dam Reservoir Area, Southeastern Washington*, 1964).

Even though Gina was distraught over the prospect of losing the ranch, she was excited about the artifacts the archaeologists were recovering (necklaces of copper beads, elk teeth, and snail shells strung on Indian hemp), and she helped with the excavations. Dooley remembers that she was so excited she would go for days at a time without eating.

I don't know how Liz felt about the digs at the time, but she now considers them a desecration. On a ride up Thornbush Creek one day, we stopped to rest, stretching our legs at the lip of a circular depression. Bertie said casually, "This is an old tepee site." Across the open hillside dotted with trees were scattered eight or nine other similar depressions. The archeologists, I later learned, had been particularly excited about this site. What they called "the most outstanding feature" was a prehistoric burial cairn containing two skeletons. Turning the page of their report, I felt a shock at seeing a photo of the remains—a man and a woman buried together, looking exposed and vulnerable curled side by side there in the dirt. I quickly flipped past the picture. The text noted that the bodies had been wrapped in tule matting, so the scientists must have peeled it back to take the picture. It seemed somehow indecent to gaze at them, as if, in the interest of science, the archaeologists had lifted the blanket on this couple's marriage bed.

While preparations were being made for the Asotin project, dams were "contemplated" at four more sites upriver from the Burns Ranch: Nez Perce, China Gardens, High Mountain Sheep, and Lower Canyon. Burns family relatives in Asotin watched as the federal government bought up property along the river. The Army Corps of Engineers moved houses, a hotel, a small hospital, and the post office away from the water's edge.

"For years, we didn't know what was going to happen to the town, whether they were going to build the dam or not," Dooley's cousin told me. From his house high on the hill, he pointed to a spot at the south edge of town. "The dam would've gone right there." At least Lower Granite Dam, ugly as it is, is in the middle of nowhere. Asotin residents would have had to live with a two-hundred-foot-high dam smack on top of them.

Sportfishermen and environmentalists—two groups who were seldom in agreement—came together in the middle 1960s to oppose the dams and to build support for the creation of the Hells Canyon National Recreation Area. After a hard battle led by the Sierra Club and Idaho senator Frank Church, the recreation area came into being in 1975—the same year that slack water reached Lewiston as the reservoir filled behind just-completed Lower Granite Dam. Although Asotin was beyond the borders of the recreation area, Congress decided to reverse its earlier decision and de-authorize the dam project. Thirteen years after getting the go-ahead, the Army Corps of Engineers were forced to abandon their plans for a dam at Asotin.

Private power companies were quick to realize that the clause prohibiting dam construction at Asotin applied only to federally funded projects. Plans for a privately built dam were put back on the drawing board, and the issue continued to be debated throughout the rest of the 1970s and most of the 1980s. Finally, in 1988, the court ruled that no how, no way, was any dam—federal or private—going to be built at Asotin. After thirty years, the threat to the ranch was finally gone. Liz was thirty-four years old. After three decades, she could sleep without dreaming of rising waters.

I walk through the orchard to say good-bye to Dooley. The last of the light leaves early from the canyon at this time of year. The sky is a seamless gray—a gray stolen from the river now in shadow. A few out-of-reach pears are faintly visible, pale bulbs that might blink on like streetlights any minute now. The cold is moving down from the mountain. From Dooley's room in the schoolhouse comes a warm yellow light and the sound of an old piano, brittle as dry leaves underfoot.

I step to the porch and stop at the sight of the old man's hunched back, swaying as his hands romp up and down the keyboard. Long underwear shows through a worn place on the back of his shirt. There's a fire in the wood stove. I stand watching him, reluctant to interrupt. But we need to get across the river before it's completely dark. I knock on the window and wave. He hollers hello and motions me in. I open the door and step into a room warm and bright and smelling of winter pears. I thank whoever there is to thank that this place, this person, this moment was not located, described, photographed, then destroyed.

Dooley walks me back to the house, where Ambrose is waiting, worn out after a long day. We talk about sturgeon. Liz told me she has often been roused out of a sound sleep by the slap of a sturgeon on the water, so loud it echoes against the cliff. It reminded me of Smokey's journal entry about being awoken by a whale's "unearthly sighs in the night" just yards from the cabin on Square Island. The sound spooked my grandmother, although she admitted it was silly to be scared of something in the water when she lay snug in bed. The thought of underwater behemoths is unsettling no matter where you are but especially so in a small boat. One day a sturgeon surfaced behind me when I was rowing a visitor across the river. I heard it blow and roll like a sea beast. The woman looked over my shoulder and yelled, "What the hell is that?" like she had caught a glimpse of old Nessie herself.

There is something I want to ask Dooley. I've read improbable accounts of ghostly white sturgeon rising out of the water, and I want to know if they are true.

"Yup, seen it myself. There was one used to come up regular when we were crossing at Ram's Head. Just curious, I guess. Always along about dusk, 'bout this time of day here. They just rise up about three to four feet out of the water, maybe ten feet away from your boat, and just eyeball you for a minute, then slip back down without a splash or anything. Scares the hell out of some people—think the devil's after them." He laughed, then was silent a minute. "It's a sight that'll stir the soul."

Coyote has disappeared. No one knows where he is anymore. The Nez Perce believe he will come again and his return will signal a change for the world. Coyote will destroy the dams, and the forests will stand again. It will be a time of reconciliation and healing. Everyone on earth—all the generations that have gone before and all that will come after—will live together as one people.

They are still waiting for Coyote.

EIGHT

Skin and Bones

The decade of the 1960s was an age enamored of plastic. We bought things that were shiny and slick and smelled like the inside of a new car. Yet there was, in the Freeman house, something that lent texture and a rich odor to our synthetic-filled lives: the furs and skins of wild animals. A bearskin covered our couch, a tanned deerhide was draped over an end table, and the skin of a baby seal lay on the piano like a deflated balloon animal. Our living room was filled with artifacts of the Far North: a totem pole, a carved wooden mask, and Eskimo toys that required coordination and a sense of timing too subtle for us suburban kids who played direct, uncomplicated games like kickball. Guests whose own homes boasted nothing more exotic than a Naugahyde sofa were inevitably drawn to the skins. The women stroked the fur with the back of their hand or rubbed it between thumb and forefinger like it was a new voile at the fabric store. The men always asked how the animals had been killed.

Certain smells are associated with particular times in our lives; the musk of wild animals was as firmly linked in my mind with the 1960s as the odor of the old kapok-filled canvas cushions on the chaise longue. It was a smell from my past and something I did not expect to encounter again as an adult.

Santa Cruz, where I spent my early adulthood, was surrounded by peaceful state parks where animals were protected, not hunted. You could watch whales and sea lions through binoculars from the beachside cliffs, or lie hidden on a sand dune to watch elephant seals on an isolated beach. It was a town full of vegetarians and animal rights activists who would no more display a sealskin in their living room than they would wear a sealskin coat. Hunting was considered outdated and barbaric, something done by beery men who probably beat their wives, too. But in Idaho, a state with no shortage of beery men (and long-suffering wives), there are many hunters like our neighbor. He lives in a split-level house in a quiet neighborhood and often spends his weekends meticulously washing and waxing his car. Yet every fall he bags a deer and hangs the carcass in his garage before taking it to the meat cutters.

With an abundance of hunters and fishermen in the inland Northwest, you don't have to look too far to find skin and bones: the small-town cafe where a moose head hangs on the wall next to a circular saw (a disturbing juxtaposition) painted with farm imagery; the shop where a taxidermist once showed Ambrose and me a freezer full of animal heads waiting to be mounted; dogs chewing on deer hooves with the distracted air of someone biting their fingernails. Even the local hardware store has bins of fur for tying your own flies: bear, coyote, beaver, and swatches of fluorescent pink rabbit perfect for a politically incorrect Barbie's winter coat.

But these glimpses of furs and bones are just that: glimpses. At the Burns Ranch, the walls of every room are lined with the skins of bear, bobcat, badger, mountain lion, deer, elk, and rattlesnake. A sheepskin is flopped over the back of the couch. A spotted cowhide with the hair still intact covers the floor. My niece, who was visiting from Irvine, California, a planned community in suburban Orange County, looked around the Burns house and then said, "I don't understand. They act as if they like animals, but then they have all these dead animals on their walls." I couldn't explain that coming to the ranch means being plunged into a world where you eat food pulled out of the ground, where if you are cold, you build a fire, where animal furs hang on the wall because they are beautiful and because they are dead.

The hollow deer hair, the coarse, tapering hair of bear, the relief map of ridges and pockets on the underside of a hide are not foreign to me, but deeply familiar. And so is the smell that permeates the house: a dry scent like scuffed dirt, underlaid with something clinging, oleaginous, the smell of a sleeping child on a hot afternoon. I was introduced to the smell and feel of animal skins by my grandmother Smokey. Over twenty years later, I was reintroduced to them by another woman named Smokey. Liz's grandfather Hiram had been called Smokey in the army because his last name was Burns. When young Liz first exhibited her tendency to fume silently until she had a sudden outburst of temper, the family started calling her Smokey, too. I found it a pleasing coincidence that these women—both of whom were to connect me to the land in a way they couldn't have foreseen—shared the same name.

Throughout my childhood, a crate arrived once or twice a year from my grandparents in Alaska. When my father pried open the box, it released a smell of wood smoke and animal musk that became the essence of the grandmother I had never met. The box might hold an oil painting of northern lights wavering over a cabin in the snow, signed with the name she had adopted in Alaska: Smokey Johnstone. Tucked in a corner might be a piece of scrimshaw or a bellikin—a good luck figure carved from a walrus tooth. Once they sent a cigar box full of duck feathers—stiff ones you could twirl between your fingers like tiny flags, others as soft and curled as baby hair. Another time my father emptied an envelope into his palm and out tumbled a handful of small furred feet—mink paws severed by a trap. But best of all were the times my father lifted out an animal skin. We touched it, smelled it, lay on it, like dogs rolling in something deliciously dead.

A reading lamp stood on the deerskin draped over the end table by my father's chair; the light warmed the hide and it felt almost alive to my touch. I did not feel sad for the deer; I knew my grandparents depended on venison to get them through the winter months on the island. I understood from my grandmother's letters that the bear had to be shot after it returned to the cabin clearing a second time and could not be scared away. I had more trouble accepting the death of the baby seal that had been pulled from its mother's belly. That the mother seal had been shot for a three-dollar bounty made its death sound like it had taken place long ago, in the Old West; but the skin was fresh, the fur that had never touched water was soft and napless, and each claw came to a perfect, unmarred point.

The paintings, carvings, and skins enlivened our suburban living room and provided a glimpse of a faraway world where things were different—a place where bears could stroll into the yard and colored lights could appear in a sky darker than anything I knew.

With one of my grandmothers in southeastern Alaska and the other so distant she might as well have lived in Nome, I had had precious little grandmothering as a child. Indeed, I had little contact of any kind with elderly people. I was envious of Liz's friendships with her neighbors in the Snake River Canyon—women in their seventies or eighties, most of them widowed. Bertie, too, had an affinity for old people. In her years of peripatetic job hopping, she has often worked around the elderly, most recently as a food server at a nursing home. When I expressed sympathy about her having to work in such a depressing environment, Bertie said, "Are you kidding? I love those old folks. You should hear them, the things they've done. I could sit and listen to them all day."

I thought often of the old woman my mother would have become. Tough as Grandma Grace, I hoped, without her bitterness. I imagined her as the critical yet supportive grandmother she would have been to my sons. I once mentioned to Liz the lack of grandmothers in my life and in my sons' lives. She said, "There's plenty of grandmothers around. You just got to look for them." For her, the grandmothers are everywhere: the voices in the river, the eyes that you feel watching you at Ram's Head, the hidden faces that peer between the leaves in her paintings of Craig Mountain.

If I had little true connection to my grandmothers, it was not for lack of trying. I asked my parents for a trip to Alaska as a college graduation gift so I could get to know Smokey, whom I had met only once. However, my paternal grandmother was to remain forever out of reach; she died six weeks before I graduated.

My frustrated attempts to cultivate a relationship with my maternal grandmother had continued all my adult life and would end with one last abortive effort. As children, my brothers and sisters and I were under the impression that our mother's parents lived far away because their visits were infrequent and treated with great formality; in reality, they lived little more than a mile from us. When Grandpa Rhodes died, his wife moved to Crestline, a resort town in the San Bernardino Mountains, to be near her son. A year later, on a visit to L.A., I borrowed my father's car and set out to see Grandma Rhodes (Grandma Grace's daughter, Dorothy) for what I was sure would be the last time.

I had my boys with me and had also brought along my nephew; ten years old at the time, he was my grandmother's eldest great-grandchild and yet she had never met him. After hours of losing our way on winding mountain roads, the boys got carsick, my nephew vomiting in the back seat. We arrived at my aunt and uncle's house at eight-thirty in the evening, long after our expected arrival for dinner. Uncle Ted called my grandmother's house across town and said we would be right over. From the other room, I listened to the one-sided conversation, my face growing hot as his voice rose in frustration. "Yes, I know it's getting close to your bedtime. . . . Can't you stay up a little later tonight? . . . They drove five hours to see you. . . . She's got the boys with her. . . . Yes, her two and Julie's son. . . . They have to leave tomorrow morning at nine. . . . Well, couldn't you get up earlier just this once?"

My siblings and I each got a small inheritance when Grandma Rhodes died a few months later. I decided to use the money to at last get to know my other grandmother, Smokey. That she had been dead for thirteen years wasn't an obstacle. I planned to visit all the places she had lived in Alaska, even going out to remote Square Island. My husband had no interest in making the trip and my kids were too young. I called my father in California and invited him to come with me. He said yes before I had even finished my sentence.

My father had suffered major complications from a liver transplant four years earlier, and although he was only sixty-two, he seemed more like eighty. He breathed heavily at the slightest exertion and tired easily. Climbing a flight of stairs was a major challenge. His hands shook from the dozens of pills, primarily immunosuppressives and steroids, he took at timed intervals throughout the day. To anyone else, he probably appeared in no shape to make a month-long trip by car, ferry, and fishing boat, but I knew he would be game for it. He had loved Alaska since he was stationed on Kodiak Island during the Korean War, and over the years he had made several trips to visit his mother on the island and, later, at the Pioneer's Home in Sitka. I explained my itinerary to him. He, too, wanted to find out more about the mother he had barely known.

We drove due north from Pullman, crossing the border into Canada in less than five hours. When we stopped for the night, my father was hardly able to climb the stairs to the motel room, but during the day, at the wheel of the car, he was his old self. He rarely ate, but when he did it was with great relish—a carton of plump cherries bought from a roadside stand along the Fraser River, a plate of pastries at a small-town bakery in British Columbia. We boarded the ferry at Prince Rupert and took it to the end of the line in Skagway, the jumping-off point for gold miners going to the Klondike.

When my grandparents had lived in Skagway, it was a town of a few hundred hardy and (by her accounts) eccentric souls served by a Canadian supply boat that arrived once a month. The number of permanent residents hasn't grown much, but every summer Skagway is inundated with tourists from the huge cruise ships that land there every day. People in the small towns throughout southeastern Alaska feel besieged by tourists, and many residents regard them with outright hostility. It is a far cry from the attitude in the years before statehood when summer visitors were seen as a welcome link to the outside world. In the early 1950s, Skagway residents used to gather to wish each boatload of tourists bon voyage. As the ship pulled away they stood on the docks, singing and waving like friendly South Sea islanders.

Beyond the boardwalk-lined streets of downtown, Skagway had changed surprisingly little. There were people who still remembered Smokey and Scotty Johnstone. We located the houses they had lived in, all still standing (some, I would wager, with much of the same furniture). The native crafts store where Smokey had worked for a time still sold beautiful bentwood boxes and soapstone carvings instead of the kitschy souvenirs we saw elsewhere in town. We had coffee with one of Smokey's friends, the proprietor of the art store where Smokey had sold her paintings.

I had contacted the present owner of Square Island and he offered to let us spend the night on the island. The forty-five-mile ride from Ketchikan in a small boat was agony for my dad. Every time the boat thwumped over a wave, he grimaced and held his stomach. "Seasick?" Jedd called over his shoulder, his hand on the wheel. I didn't attempt to explain over the roar of the motor that surgery after surgery had left my father's innards a mass of scar tissue.

In Alaska, especially in the bush, things tend to stay put. Once something was hauled out to the island, it usually was there for good. We found that much of the furniture from my grandparents' day was still in the cabin. The windup record player still sat in a corner. My father would have liked to take it home with him, but Jedd didn't want to part with it. He had no problem with me taking the flour sifter my grandmother had used to make their weekly batch of bread.

"What color would you say that is?" my father asked, touching a finger to the low ceiling.

"Black."

He wiped away a layer of soot, revealing a paler undercoat. "I painted this white when I was here in '64. Looks like it hasn't been painted since."

A well-thumbed notebook lay open on the table. It was the same kind of weather log my grandparents used to keep to record the daily temperature and wind speeds. I flipped through the pages and read entries by Jedd's friends and far-flung neighbors who dropped in to visit. There was also an occasional note from a grateful stranger who had found the cabin open in winter.

Backcountry cabins throughout the West are traditionally left open and stocked with food to provide emergency food and shelter for travelers. However, the practice is less widespread than it once was because many cabin owners feel that they can no longer depend on visitors to follow the unwritten rules: refill the wood box; stop by with a few cans of food next time you're in the area; above all, respect the owner's personal property. Some owners, tired of replacing stolen tools and chopping firewood to replace that used by their uninvited guests, have taken to locking their cabins. Like the Burns family, who leave Percival's house on Craig Mountain open and stocked with food, Jedd thinks it's worth the risk. In the event of bad weather or an accident, the cabin may mean the difference between life and death. Knowing that there are many hunters and fishermen still in the area when he goes back home in the fall, Jedd leaves the door unlocked and the log open on the table.

My father sat on the porch to read and wait for Jedd to return from checking his crab pots. I hiked a short way along the few trails, but they were badly overgrown. The undergrowth was so thick that just a few feet into the forest, the shore became invisible. It would be easy to get lost here. I had grown used to the dry, open spaces of the inland Northwest. The rampant vegetation and huge trees that blotted out the sky made me claustrophobic.

I walked out onto the point where, to Smokey's great relief, she had caught sight of Scotty coming home in the dusk after a bad leak in a borrowed boat had left him stranded on the other side of the island. Across the water, I could see my father sitting on the porch next to a corner of the garden fence. It was the same corner where Scotty had found an antler tangled one morning after a buck had tried to leap the fence from the porch, less than seven feet from where he was sitting reading by lamplight.

The evening was endless. The light stretched on and on and there was no discernible drop in temperature. Smokey, too, had noted how odd this was, accustomed as we both were to the abrupt contrast between day and night in Southern California. In the beach cities, fog rolled in at dusk like steam from the red hot sun plunging into the water on the horizon.

My father and I built a fire and put water on to boil. Soon Jedd anchored his boat offshore and paddled a dinghy to shore. He showed us how to tear apart live crabs with our hands and throw them into the boiling water. We sat on the porch, eating. My father took a long time to eat his crab, digging the meat out with shaking hands, chewing slowly and stopping to gaze out at the bay. He looked at everything with the fierce attention of someone who knows that this will have to last him.

There was a bald eagle's nest at the top of a Sitka spruce a few yards from where we sat. Earlier that afternoon we had used binoculars to see the punk-haired heads of the chicks as they greeted the parent bringing them a fish. Both male and female flew back and forth to the nest, swooping so low over our heads that we could feel the rush of air beneath their wings. The Johnstones would not have shared my enthusiasm for the bald eagles. Sea lions, killer whales, and eagles were despised because they ate the same fish Alaskan fishermen depended on for their livelihood. As predators, they also scared away the more appealing animals, such as otters, dolphins, and ducks, that my grandparents enjoyed watching.

As we sat on the porch looking out over the water, I kept an eye out for killer whales like we had seen from the ferry. My grandparents had often seen killer whales here, but their attitude toward them was anything but welcoming. Smokey always alerted Scotty when she saw killer whales or sea lions in the bay. He would interrupt whatever he was doing to grab his gun and shoot from the porch. Scotty did not aim to kill the whale; he hoped only to scare the pod out of the bay. My grandmother admitted it made her feel slightly sick to hear the bullet thunking into the dorsal fin—a sound that carried easily across the water. The thought of needlessly wounding such an animal makes me feel slightly sick, too.

The Johnstones took a similar approach to eagles near the cabin. Smokey used the eagles for target practice, aiming not at them but at a nearby branch; when she hit her mark, the bark flew, and the eagle lifted into the air, she congratulated herself on ridding the island of another pest.

After skinning a seal, Scotty used to set the nose on the windowsill to dry. I had half-wanted him to send us an envelope of those black noses; I imagined them spilling into my father's hand like licorice buttons. But the state bounty was payable upon receipt of the nose, and Scotty guarded them carefully. The local squirrels had developed a taste for seal noses and often tried to swipe them from the windowsill.

I had read my grandmother's thoughts on predators such as the wolves and bears she watched on the mainland a third of a mile away. I didn't accept her characterization of wolves as savage and bloodthirsty, yet I recognized that it was a reflection of the time and place. She was a well-educated professional woman, but when she moved to the Alaskan bush, she adopted wholeheartedly the local prejudices about wildlife. It was part of her effort to remake herself as an Alaskan. She had married a true sourdough with an encyclopedic knowledge of the bush. She was a cheechako, a newcomer, and the title chafed. She struggled to learn how to shoot a gun, skin and prepare game, cook on a wood stove—things that Alaskans seem able to do as soon as they're out of diapers, the whole state populated by junior Pecos Bills out whomping wildcats at the age of three.

The Johnstones spent their last years in the Pioneer's Home in Sitka, a free rest home for old-timers who've been in the state for twenty-five years or more. Smokey took pride in the fact that she qualified as a "pioneer." She felt that it was a title she had earned.

My father and I spent the night in the wanigan behind the main cabin. The wanigan is a one-room cabin Scotty built to provide emergency shelter if the main cabin burned down. The main cabin is perhaps three times as big as the wanigan, but it is still only the size of a two-car garage. Many people would go stir-crazy in such a small cabin on an isolated island, but it seldom bothered Smokey. By choice, she went to town only about twice a year and could never wait to get back to the island; the beds were too soft, street noises too loud, and the sidewalks bone-achingly hard.

I lay awake on the plywood bunk, shifting on the thin foam cushion, long after my father was snoring in the other bunk. All day I had looked for my grandmother—in the overgrown garden with the towering hollyhocks she had planted, at the bench where she used to sit soaking her feet in the creek, by the wood stove that had been such a trial to her. I had been standing on the beach wondering where she was when I looked down and saw at my feet a flat rock with the word "BOO" in raised white letters. The hair stood up on the back of my neck. I picked up the rock and laughed to see the word was just a cluster of barnacles. Well, I thought, if this was Smokey's doing, it was a pretty good joke.

My father and I agreed that we didn't get much sense of Smokey on the island. We had felt closer to her on the dirt streets of Skagway, where we half-expected to come around a corner and run into her. The Alaskan bush was Scotty's world. He was utterly at home here, and almost everything visible, aside from the main cabin, was the result of his handiwork. Smokey had entered this landscape, observed it closely, and described it in her journals, but she had not left much of a mark. But the smell I associated with her was here, recognizable the instant I stepped into the cabin—wood smoke and animal skins. It was complicated with other scents: salt air, drying boots, mildew, pine, something like tar. I tried to fix the smell in my memory. I could bring home photos but not the whoosh of the eagle's wings a dozen feet over my head, or the smell that was my grandmother. Like my father, I had to absorb everything I could on Square Island. This would have to last me, too. I would probably not be back, and if I did return, the island would have changed irrevocably.

Alaskans have been protesting for years that almost the entire state is locked up in government lands; in response the state is making more land available for private ownership. Square Island and the southern half of its spacious bay were selected for "remote disposal"—meaning sale to private citizens. The term reflects the state's utilitarian approach to its lands. Indeed, the plan is to subdivide the island and sell the lots for "community expansion." Sounds like urban sprawl to me, yet the island is forty-five miles from the nearest town. In Alaskan terms, that's just a hop, skip, and a jump, and Square Island is considered ideally situated for weekend cabins.

I later learned that several years after our visit—at the time the land was transferred to state ownership—Jedd was allowed to buy the small cove where the Johnstones had lived. I heaved a sigh of relief when I found out that Jedd was able to hold on to the cabins. If the land had remained part of the Tongass National Forest, the "improvements" would have reverted to federal ownership after Jedd's death, with the likely result that the cabins would have been burned.

It was reassuring to me that someone who has a deep love of the island and an appreciation for its history will be there watching over things. Jedd's watch may not be an easy one. An alternate fate may await Square Island: if the university includes it in their request for state lands, the island may change hands once again. University property has one purpose: to bring in funds for higher education. The entire island could be sold, logged, or commercialized—all except for Jedd's tiny one-acre lot.

I can imagine the response of my bantamweight Marine Corps grandfather to the government coming in and trying to burn down his hand-built wanigan. Like many westerners, he was strongly patriotic but also keenly suspicious of the federal government. The Johnstones had always kept the place shipshape, with a flag flying overhead and the beach raked clean of kelp (the island equivalent of shoveling one's walk). Scotty would undoubtedly be appalled at the present condition of his small island fiefdom.

There was an unmistakable air of decay enveloping the small clearing. Logs were jackstrawed on the beach; Scotty would have fended them off with a pike pole or winched them onto blocks and cut them for firewood. The log float once anchored in the cove was now high and dry, carried ashore by a long-ago winter storm.

The Johnstones had been particularly proud of their cache: a traditional Alaska log structure about the size of a doghouse on stilts. Now the door was hanging by one hinge, and instead of a side of venison or halibut inside, there was a pile of rusty tin cans. The old trapper's cabin in a nearby cove, still standing in my grandparents' time, had collapsed and was almost hidden by a jungle of vegetation. Jedd spends only a few weeks or, at most, a few months on the island each year. When he's not fishing, he spends most of his time doing repairs and cutting the brush that threatens to reclaim the clearing. He figures it will take several years to make up for over a decade of neglect.

Before we left the next morning, my father stood on the beach looking across the water at the misty Rock Candy Mountains in the distance. He shook his head, saying, "This place is a dream, just a dream." When my father got out of the navy, he had wanted to stay in Alaska, but my mother firmly rejected the idea. She had lost a baby in Kodiak and the North meant nothing but heartache. She wanted to raise her children somewhere civilized and safe.

It saddened me that my father had been able to resign himself to living in Los Angeles only by convincing himself Alaska was an unattainable dream. That was what had driven me to leave Southern California at age nineteen: the fear that eventually I, too, would come to think of the city as the real world. Here among the remains of my grandparents' lives, I felt what I never could in L.A.: a sense of time passing, of things aging and falling apart, revealing something solid and changeless at the core. "No, Dad," I said, "*This* is what is real."

My father and I stopped in Ketchikan before boarding the final ferry home. Walking down the street, gazing about with that peculiar concentrated air, he suddenly fell flat on his face. He spent the rest of the trip sitting on his bunk on the ferry, his arm encased in a huge cast, looking out the window or reading. His skin was a strange purplish-brown, his wide shoulders nothing more than a yoke of bonework. I was grateful for these weeks I was able spend with my father. I had lived hundreds of miles away from my parents for virtually my entire adulthood. And I can recall only one occasion in my childhood when my father and I had gone somewhere together: a father-daughter dinner at Girl Scouts. Every moment with him was precious. I didn't know how long he had left.

The trip had given us a good sense of what Smokey's life here had been like, but we were no closer to understanding her as a mother or a grandmother. We talked about her at length, speculating about why she had left her son at boarding school, why she hadn't stayed in the newspaper business, why she had come to Alaska. My father was sitting on the edge of the bunk, shirtless, his stomach crisscrossed with staple marks and the broad Y-shaped scar that marks a liver transplant patient. It seemed that after sixty-three years on earth, the scars of my father's childhood had worked their way to the surface. "Maybe she went to Alaska to get away from me," he said in a voice like a lost child.

Less than seven months after our trip to Alaska, my father died of lymphoma. It is a form of cancer that grows quickly and easily in transplant patients taking immunosuppressive drugs.

After his own lonely years at boarding school, my father had welcomed the commotion seven children brought to his adult life. He was never happier than when we romped through the house, wrestling and pulling each other down the stairs. "Keep it down to a dull roar," he would say, drifting off to sleep amid shouts, loud thumps, and laughter. It was no surprise, then, that my father did not die with all of us crowded around the bed in silence. The nurses assured us his heart would stop within a matter of minutes after they turned off the life support machines. It beat for hours. His blood pressure dropped so low the nurses called their colleagues into the room to marvel at the numbers.

We had told my father the night before that his oldest friend, Jerry, was on his way from Japan, where he was living at the time. Jerry is godfather to six of the seven of us (our youngest sister wasn't born at the time of our group baptism). My father's heart stopped beating just as Jerry's plane touched down at LAX; his kids were in safe hands. We were sprawled all over the waiting room—talking, braiding each other's hair, eating See's candy, playing cards. My eldest sister had gone down to the cafeteria, my younger brother was stretched out on the floor, his forearm over his eyes. Amid the hubbub of a large family, my father slipped easily away.

From his belongings, I chose a painting Smokey had done of the abandoned docks at Skagway, the pilings jutting toward the sky. I also took a doeskin she had sent us many years ago. In my father's house I pressed the soft skin to my face. The smell of smoke and skin had become not only the smell of Alaska and the grandmother I never knew, but also of Idaho and the grandmothers there I have barely begun to know, and I understood why I feel most alive in places that smell of death.

NINE

The June Drop

One warm day in mid-June, we walked through a boggy pasture to one of the best apricot trees on the Burns Ranch. Ripe fruit the color of California poppies lay scattered in the grass. Liz looked with satisfaction at all the apricots and said, "The old-timers call a windfall like this the June drop." Ambrose eagerly reached for the fruit over his head, but Liz told him to start with the ones on the ground so they wouldn't go to waste. We turned over each promising-looking apricot to see if the underside was bruised, rotten, or crawling with wasps. When we found a good one—maybe just a slight brown spot on one side—it went into the canvas bucket. Ambrose, on his knees in the grass, complained, "This isn't picking apricots, this is picking *up* apricots."

I understood Ambrose's impulse to pick the perfect fruit hanging from the branches. When I was about his age, we went every year to help my great-uncle (Grace's younger brother) pick oranges at his grove in Orange County. There we tasted fruit we never saw in the supermarket, old-fashioned fruit like persimmons, figs, and loquats. We slipped the skins off dusky, fragrant Concord grapes. Each of us claimed an orange tree heavy with fruit and raced to see who could pick the fastest. My big brother did pull-ups on the branches, showing off. My little brother pelted us with green oranges.

I often stopped to rest at the top of a tree, pulling a kumquat from my pocket to nibble the sweet rind while I looked out over the farm and the roofs of the tract houses that grew closer every year. Uncle Herman took great pleasure in his orchard and garden. He had worked for several decades at a dry goods market, starting as a young man delivering groceries by horse cart. I imagined that all those years of handling bags of cornmeal, baking soda, saltines, and bolts of cloth had made him long for the feel of a good round plum. Inside the house that smelled of mildew and old gravy, my father would wash his hands after fixing an irrigation pipe, and my mother would shake her head sadly over the photo of Aunt Mary that Uncle Herman always brought out, as if to include his long-dead wife in the day as well.

With no children of his own to take over the farm, my uncle eventually sold out to the developers who had been hounding him for years. By the time he moved into a nursing home, the farm was an island surrounded by Spanish ranch-style subdivisions. Soon the acres of bright tile roofs would be the only thing orange about Orange County.

Liz and I carried the full buckets back to the ranch house and washed the fruit in the sink. I picked up a handful of apricots and cradled the tiny baby bottoms for a moment, all plump and warm, before setting them on the cutting board. We settled into the rhythm of slicing the fruit and tossing pits and bruised sections into an old coffee can. In the summer heat of the kitchen, the smell of apricots was overpowering—a pungent mixture of oranges, honeysuckle, and vanilla bean. Outside a dirty window, hummingbirds buzzed at the feeder. Inside, the parakeets talked companionably in their cage in the corner. My hands worked without thought, and I found myself looking over a recipe for rabbit stew on the wall, making mental notes as if I actually planned to make the dish someday.

At my side, Liz gave a shiver and paused for a moment. Picking up another apricot, she said, "The dead moms again." My mother had died a year before we moved to Idaho. Liz's mother had died two years earlier. That both our mothers had died of cancer relatively recently was one thing that had brought us together. Both Liz and I experienced flashbacks of our mothers' deaths that appeared suddenly with an almost hallucinogenic intensity. The sight of a dead bird—its wing angled awkwardly across its body the way my mother used to hold her arm protectively over her missing breast—was enough to make me weepy all day. Liz called those moments of searingly painful remembrance "the dead moms."

"You remind me of my mom—the way the two of us used to stand here canning fruit every summer. After she died, whenever I tried to make jam, I'd just sit in the kitchen and bawl." There are stacks of old canning jars in a corner of almost every building on the ranch as well as in the dump, but Liz doesn't can much anymore. She's had enough days of standing over a steaming kettle in the hottest part of the summer. It's faster and easier to dry the fruit in a dehydrator.

A few days earlier I had gone down to the cellar for the first time. I was looking for a hummingbird feeder to replace the one lost in a winter windstorm. Kicking open the stuck door, I felt around for a light switch. Jars glowed a sulfuric orange in the light from the doorway. On the dirt floor were bits of old bones and the remains of a broken light bulb. Huge vinegar bottles with glass stoppers stood empty against the bulging rock walls. As my eyes adjusted, I saw warped shelves lined with full jars of cherries, tomatoes, corn, peaches, and green beans. The lids were rusty, the labels faded. Flaccid pickle slices floated among seaweed-like sprigs of dill. I took a jar of the orange stuff (pumpkin? squash?) into the full sunlight in the doorway, but the label's scratchy handwriting was too faint to read. The hummingbird feeder, a jaunty thing shaped like a hot air balloon, hung from a wire stretched across the low ceiling. I took it upstairs, scrubbed it, and filled it with a sugar solution. The ruby-throated hummingbirds, which Liz hadn't seen for weeks, showed up the next day.

Liz and I talked about the horse-pack trip we had been planning to take up to Selway Falls. But, Liz and I agreed, it wasn't a good time to take a long hike into the wilderness. Something had come up that screamed "the dead moms" loudest of all. Bertie had found a lump in her breast. She had already undergone a biopsy, but it would be days, perhaps weeks, before she would get the results. Maybe, we thought, a short pack trip up into the mountains behind the ranch would help take her mind off things.

I knew Ambrose would be game for a hike up the canyon. He was hoping to see a bear. Returning from a trip to Yellowstone the year before, we had stopped along the Lolo Highway where a suspension bridge leads across the Lochsa River into the Selway-Bitterroot Wilderness. It was the closest Ambrose had come to true wilderness, and he was eager to set foot in it. Emlyn had just informed him that bridges work exactly like tunnels: if you hold your breath until you get to the other side, your wish comes true. Ambrose took a deep breath and raced across the swinging bridge. He stood at the edge of the forest, peering into the trees expectantly. We ventured a short way up one of the narrow trails, the tall trees closing behind us so that the river, just steps away, disappeared from sight. We turned back at a fork in the trail. Ambrose walked more and more slowly until, in the middle of the bridge, he stopped altogether, staring down into the rushing, rock-strewn Lochsa.

Ambrose didn't talk until he was almost three, and in those silent years I had learned to read his face well. As he stood over the river with the wind blowing the hair back from his pale forehead, there was a tender, wounded look about his lips, as if life had just given him a sharp slap. He was deeply disappointed about something.

Craig Mountain from the top of Rice Creek Grade, 1952. Photo by Kyle Laughlin. 99-H-057-04.
Historical Photograph Collection, University of Idaho Library, Moscow, Idaho.

In the car, puzzled but still not doubting his older brother's word, he
said, "I ran all the way across without taking a breath, but I didn't see a
bear. Why didn't I get my wish?"

A few days after our conversation in the kitchen, we started out for a
two-day trip on Craig Mountain. We followed the course of Thornbush
Creek as it led up the canyon. Ambrose skipped ahead, fell behind, and
detoured into the woods whenever something caught his eye. Doglike, he
traveled twice as far as the rest of us to cover the same distance. Despite an
old knee injury from the time a horse rolled on her, Liz kept up a steady
pace for hours. Every time the road crossed the creek, she walked through
the water without breaking stride. Bertie, leading the horses, did the same.

I hopped from rock to rock, trying to keep my feet dry. On the third or fourth crossing, I slipped on a patch of moss and fell knee deep in the water, scraping my hand on a sharp rock. I had to admit that Liz and Bertie's method made sense. When following cattle up into the hills, they couldn't risk an injury over something as inconsequential as wet feet. My damp pant legs flapped with each step, and I looked enviously at the army-style pants tucked neatly into Liz's boots. Good for keeping out ticks, she said. I reminded myself to check my legs next time we stopped.

Along the cliffs on either side of us, Liz pointed out various caves—some in which her mother had found Indian artifacts, others that were home to great horned owls. When we passed a large sideways slit of a cave that Liz could tell us nothing about, we decided to stop and explore. We scrambled up the talus slope, our feet causing tiny rock slides behind us. I kept Ambrose directly in front of me in case he fell, but he was more nimble-footed than I; he moved quickly up the hillside and waited on a narrow ledge. I struggled upward, clutching at handfuls of grass and ropy weeds that pulled out by the roots.

Liz led us single-file along the ledge to a deep overhang going back about twenty feet. A low wall, angled toward the front of the cave, led to a circle made of stones. Liz studied the configuration and decided it was the remains of a sweat lodge. Once she pointed it out, I could see it clearly. The low rock wall, probably lined with mud, had channeled water from a small seep at the back of the cave. The ruins, embedded in the ground and covered with fine dirt, looked hundreds, perhaps thousands of years old. But Liz reminded us that this was the Nez Perce's land as recently as a century ago; they might have built the sweat lodge. Today, the border of their reservation is less than five miles from the cave.

We sat down at the cave entrance, sweating from the hard climb, and felt the cool wind blowing up from the canyon floor. The Indians, whoever they were, must have squatted here at the edge of the cliff, steam rising from their bodies. I looked at Ambrose staring peacefully down at the trees below, enjoying the feeling of height our vantage gave him. This is what I wanted for my son. I couldn't promise him a bear; I couldn't give him the true wilderness he longed for. But I could bring him to places where he could experience moments like this. My eyes drifted to the ceiling of the cave, and there I saw a small red handprint no bigger than my son's hand resting like a blessing above his head.

We had gotten a late start, and our progress had slowed after we left the flat canyon floor and followed the steep, rutted road into the mountains. It was early evening by the time we arrived at Percival's house in a small cul-de-sac above Thornbush Creek. Ambrose was drooping with exhaustion.

We sat down by the side of the road below the house while Liz took the horses to scout out a place to camp in a meadow up ahead. We waited in silence, glad to be off our feet.

Then we heard and felt a pounding on the road and the Appaloosa bolted around the curve of the road. His pack was missing. He careered past us, his hooves sending stones out behind him with a cracking sound, and disappeared between the darkening trees. Bertie automatically lit out after him, whistling and calling his name. I would have just let the horse go; it might be miles before she caught up with him.

I called for Liz but there was no answer. I was reluctant to leave Ambrose alone in the dim light while I went up to the meadow. I was sure Liz had the situation under control and would appear any minute.

Over a half an hour passed before we saw Liz come around the curve leading the other horse.

"Where the hell is Bertie?" were the first words out of her mouth.

I explained that she had gone after Gramps.

"Chasin' the damn horse. Didn't bother to find out if I was all right first—that stupid animal could've brained me."

I was ashamed that the same thought hadn't occurred to me. I was so sure that Liz could always take care of herself.

"We'll have to camp here. There's a dead cow in the meadow. When Gramps smelled it, he just went crazy. I couldn't hold him."

Ambrose asked, "What took you so long?"

"I had to repack everything so Dixie could carry both saddle bags. Come on," she said, heading up onto a grassy hill, "you look like you could use some dinner."

I helped Liz unload the saddle bags and unpack the camping supplies. When it sunk in that we were spending the night there, Ambrose broke down and wailed, "You mean we hiked all this way just to camp next to a house?"

Sulking, he dragged his sleeping bag a dozen yards away from us and lay down in it, chewing a granola bar and looking up at the dark circle of mountains outlined against the still bright sky. He had hoped to hike deep into the wilderness where he might catch a glimpse of a bear.

The pack had landed squarely on the compartment holding the cooking utensils. Before Liz and I could prepare dinner over the propane stove, we had to hammer the aluminum plates and cooking pots back into shape. About the time dinner was ready, Bertie showed up on the winded Appaloosa. Liz chewed her out about her lapse in judgment. Horses are expendable; people aren't. Bertie was apologetic: "I just didn't think."

I understood that ranchers need to be able to rely on each other in the dangerous situations that often crop up, but my sympathies were with Bertie. Like me, she had been raised in the city and dreamed of the mountains and of having land of her own someday. (The difference between our dreams was that hers contained horses.) Working on this Idaho ranch is the closest she'll ever get, while I consider myself fortunate to be able to visit once in a while. But even after years of branding calves, herding cows, and working with the horses, Bertie is still learning. The knowledge came slowly because much of the instruction was unspoken. She was to learn by example, and it was up to her to make sense of it all. I knew that this was how Liz had learned the ranch business from her father. I thought it was just the cowboy way, but Liz said that when Pop was showing a guy how to do something, he would talk his ear off, explaining things. I thought, not for the first time, that Liz's life would have been easier if she had been born a boy.

As an only child, Liz knew exactly what was expected of her. In her final year of college, she received word that something was seriously wrong with her mother. It had been weeks since she had left the house, and, Dooley said, for days now she hadn't gotten up from the couch. She refused to go to town to see a doctor. Liz flew home immediately. As soon as she saw her feisty, energetic mother curled in a fetal position on the couch, she knew that she was dying. Liz cajoled her into going to the hospital. Weak and in pain, Gina agreed. They lifted her from the couch and carried her down to the beach. Settling her gently in the boat, Liz rowed her mother across the river for the last time.

Gina died at the hospital two months later without ever having come home. There was no question of Liz going back to college. She was needed on the ranch. Her father was in his midsixties and the ranch work was too much for him to handle alone; there were money problems, and much of Dooley's motivation was gone. As the years went by, Liz's father handed over more and more of the day-to-day responsibilities to her. She became indispensable, and her youthful plans—to finish her education, become a therapist, travel—were swallowed up by the never-ending demands of the ranch.

I once asked Liz what she thought she would be doing if she had gotten her degree in psychology. She responded quickly, dismissively: "I probably wouldn't have graduated anyway. I was always getting into scraps with the teachers. They thought I was too opinionated."

"But what if you had graduated?"

"I'd probably be trying to comfort unhappy people in some little town."

When Liz was working outdoors, she often spoke in an abrupt, commanding tone. There was work to do and she wanted it done her way. But when she was not involved in the strenuous business of keeping the ranch going, she showed a softer, deeply empathetic side. She went about what she was doing—applying mink oil to her boots, doing delicate beadwork—to all appearances not hearing a word you said, until you stopped talking and she answered with an understanding comment.

As a college student, she had worked for a domestic abuse hotline, and I asked her once how she had distinguished the whiners from the people with real problems. She said, "Well, some people are just looking to stir things up—give themselves a crisis high. Then there's others who just want to complain. They say, 'Yes, but . . .' to everything you suggest. But the ones who're really in trouble—when you offer them help, they grab onto it like you've thrown them a life preserver." She still gets late-night calls from friends, acquaintances, sometimes friends-of-friends, who know she is a good person to talk to when you have troubles. More than once, sleeping in the front bedroom, I've awoken to the sound of her phone ringing. Between long silences, I heard Liz's soft, calming voice. I thought about the desperate person in town and Liz here in the canyon, and I fell asleep trying to sort out which one was the voice in the wilderness.

The next morning I woke to the sun coming over the east ridge and lighting up the tops of trees on the western slope. Ambrose was still asleep, so I walked down to the old house to look around. After Percival's wife died, he moved away from the rest of the family down by the river and built the house in this remote valley. It is a lovely, peaceful spot—an open hillside sloping down to the creek, surrounded by steep mountains dotted with huge ponderosa pines. Each time I visit the house there are small changes: another section of the porch roof caved in; a mound of dead bees (killed by the first frost) beneath an upstairs window; a can of sardines missing from the small stack of canned food on the kitchen table.

The house is left open for hunters or others in the area who might need shelter or food in bad weather. I would have to be pretty desperate to spend the night in this house. From high on the mountain, it is small and white, as picturesque as an old church. From close up, it is slightly less appealing, as the condition of the house becomes apparent. And stepping inside it is downright creepy, with the bone-chilling cold of the long abandoned. The overstuffed chairs in the living room are an explosion of springs and padding. The shelves have collapsed, spilling mildewed books onto the floor. In an upstairs bedroom, lace curtains still frame a view of the ridge. The long enclosed porch at the back of the house has collapsed like an undercut bank.

Somewhere out there, hidden under the trees, are Percival's and Odalie's graves. Liz is reluctant to take me there. She seems to have an aversion to graves, while I am drawn to them. Dooley says that some of the young men in the family who revere Percival and his self-reliant way of life talk wishfully about being buried up here in the hills with the old man. But we can't always choose where we will end up. My mother and father share a single gravesite in a crowded cemetery. Our small family plot is now full and no other family members can be buried there. The once quiet street has become a popular shortcut from the beach to the freeway. It bothers me that yards from my parents' grave, commuters pass by juggling their coffee and cell phones, unaware.

When I miss my parents, I go out to one of the graveyards that dot hilltops throughout the Palouse. Prairie grasses that have mostly vanished from the rest of the county still flourish on these islands of unplowed land. Worn limestone markers—a dozen, maybe two—lean at precarious angles, and a horned lark sings. On my first trip to Los Angeles after my mother died, I went to visit her grave. Arriving in the late afternoon, I was shocked to find the cemetery gates locked. It had never occurred to me that I would not be able to go to my mother. I stood gripping the bars of the gate, sobbing like an abandoned child.

I heard the outhouse door bang shut. Looking out the window, I saw Bertie kick at something beneath one of the gnarled apple trees in the yard. Later, as we fixed breakfast, she mentioned to Liz that she had seen some bear scat beneath the trees.

"Bears?" Ambrose said, suddenly all ears.

"Yeah," said Liz, "They like to come down here and eat the apples."

"Why didn't you say there were bears around here!" he shouted, taking off at a run. With all the commotion he was making, I was sure any bears would be long gone by the time he got down there. I was just glad Liz hadn't mentioned bears the night before when we crawled into our sleeping bags under the stars.

After packing the horses, we hiked up to a waterfall on Thornbush Creek. The water coursed down the mossy face of a cliff, splashing onto the rocks below and rising in a cool mist. Later in the summer, the temperature in the canyon can hover close to one hundred for weeks at a time. Old-timers claimed the canyon got so hot that the Snake River boiled at the edges. When the heat becomes unbearable, Liz makes the long trek to the waterfall and sits for hours in the shallow pool at the base of the cliff. My own response to soaring temperatures would be to head to town to find an air-conditioned library. But Liz, like her neighbors up and down the river, has learned to cope with the heat, finding ways to get a break from it without having to leave the canyon.

Ambrose made his way across the slippery rocks to stand, for a brief chill moment, beneath the waterfall. Liz sat on a rock braiding her hair, which had been pulled loose by brambles along the overgrown trail. Bertie leaned against the trunk of a tree and spoke with a shy pleasure about an upcoming visit from her parents. They had postponed a trip from California many times because of her father's poor health, but this time it looked as if they were going to make it. Although she wasn't close to her devout Mormon parents, Bertie kept in touch by phone. She was looking forward to showing them around the ranch. She hadn't told them about the tests she had undergone and didn't plan to until she knew the prognosis. If she had to go in for surgery, they would have to call off their trip yet again.

I was married and pregnant when my mother wrote me, in her usual cryptic way, that "all was not right." I had to find out from my sisters exactly what was wrong. Mom had been showering when she found a lump in her breast. We phoned back and forth; the more we talked, the more our fear grew. We had all known women who had survived breast cancer, including our godmother, Betty, who had undergone a mastectomy back in the 1960s when the standard procedure was to take out a huge, disfiguring section of the chest muscles and upper arm. It was the 1980s; we thought, surely this can be taken care of, and with less trauma than Betty had gone through. It was not, then, the appearance of breast cancer that frightened us, but the contrary way in which our mother was bound to respond.

My mother thought all doctors were quacks. In the early 1970s she had joined a nutrition society that met weekly to discuss healthy eating. Each week she came home eager to try out yet another recipe from the books of health food guru Adelle Davis. Certain that all her Betty Crocker cooking over the years had ruined our health, my mother was determined to instill good eating habits in us in the short time before we left home. Overnight we went from Wonder Bread to dense loaves of bread made with soy flour and millet. Instead of brownies in our school lunches, we got slabs of oily, black gingerbread. For Sunday dinner, our favorite pot roast was replaced with bell peppers stuffed with brown rice and peanut butter. She often refused to tell us exactly what was in the dish we were eating. "Just eat it," she said. I had constant stomachaches and lost weight. When I left for college, she sent me care packages of unsulfured dried apricots, brown and curled like little wizened ears, and granola bars packed with every nutrient known to mankind. Tucked into the corners of each care package were envelopes of instant soup that cooked up into a strange green broth teeming with bits of algaelike matter. Got enough primordial soup, Mom. Send popcorn.

One of the prominent members of the nutrition society was my mother's chiropractor, a woman who drove a yellow Mercedes convertible and talked rapturously about high colonics. Like the rest of the nutrition group, she did not take a mainstream, school dietician approach to nutrition. The group was rabidly anti–medical establishment. If the American Medical Association endorsed a particular procedure, that was reason enough to reject it. My mom had never even had a mammogram because radiation, even a minute dose, was one of the no-nos. If she swallowed her pride and turned to the nutritionists for help, we knew she would follow their advice about treatment even if it killed her.

Betty called my mother to urge her to have the surgery immediately. Our godmother described how she had come to terms with the loss of her breast; she had gotten twenty extra years out of it so far and felt it had been a good trade-off. My mother was so angry that she broke off all contact with Betty, her oldest friend. "She doesn't understand," she said bitterly. What Betty didn't understand—what none of us in the family understood—was my mother's sense of shame. She was going to have to admit to her friends that she had failed (in some unfathomable way) to improve her life through healthy eating. When I reminded her that Adelle Davis had died of stomach cancer, Mom snapped that Adelle attributed her cancer to the poor food she had eaten in the cafeteria in her college days. Sure, Mom, I said, thinking of my own relief at leaving for college, where the cafeteria food, although not terribly appetizing, was at least recognizable.

My father, an engineer who designed logic systems for Xerox, was infuriated at my mother's "illogical" response. He was a troubleshooter, used to fixing anything from broken toys to computer hardware. To him, the problem was obvious, the solution simple: cut the damn thing off. But after thirty years of marriage, Dad knew there was no point in badgering his wife once she had made up her mind.

With the exception of one sister who felt that we should support my mother in whatever course she chose, my siblings and I shared my father's opinion of what he called "voodoo medicine." We tried, by phone and letter, to talk my mom into seeing an oncologist, but she just kept insisting we give her some time "to do it my way."

My mother embarked on a long lettuce juice fast. A simple cleansing of the body, she explained, as if that would do the trick. When the fast failed to curb the growth of the tumor, she tried coffee enemas. Something about stimulating the immune system. Weeks, then months passed as she experimented with various treatments. We begged her to give up and see a doctor; she refused and made two trips to Mexico for laetrile treatments. In Tijuana, whole clinics were filled with cancer patients desperate enough to try an experimental medication made from apricot pits that was illegal in the United States. We knew that if Mom was resorting to laetrile, she must be close to admitting defeat.

My dad called a family meeting. My brother and I drove down from northern California; one sister flew in from Colorado, another from Idaho. The others arrived from various Los Angeles suburbs. When she was getting dressed for the gathering, my mom called one of my older sisters and me into her bedroom. She wanted us to see what the cancer had done to her. She removed her bra and stood silently. The only sound was our simultaneous intake of breath as my sister and I saw the misshapen mass that had been our mother's breast. My sister gently (and, I thought, bravely) touched it. "It's hard as a rock," she said. In a few short months the lump had grown so fast that the entire breast had ossified. I hugged Mom gingerly around the shoulders. "It's not you anymore, Mom." It was something ugly and dead. I couldn't see how she could stand to have it attached to her body for one minute longer. She gave a convulsive sob and said, "I just want it back the way it was."

At the family meeting, we begged, bargained ("We'll send you to Europe!"), and argued with Mom about a mastectomy. We were all in tears by the time she finally, resentfully, agreed.

The next day my mother insisted on taking my sisters and me shopping at an upscale mall in Palos Verdes Estates. She loaded us up with presents and bought us double-dip cones at Baskin Robbins, just like the old days before she had outlawed ice cream in favor of something called Soya Cream. I wanted to shout at her, "Stop pretending everything is all right!" I went into the Macy's restroom and watched blood dripping from between my legs. I thought, "If I lose my mother *and* this baby, I'll go crazy." I called my husband and he begged me to come home and rest.

Emlyn was born six weeks premature, small but healthy. My mother's cancer continued to advance. I was staying at my parents' house on the day my mother was informed that the cancer had metastasized to her bones. My mother locked herself in the bedroom, raging and crying. There wasn't anything I could do. My friend Moana offered to take me somewhere, anywhere. As we got into the car, a man who had been watering his lawn walked slowly across the street. He cocked his head, trying to get a fix on where the sound was coming from.

"You got a problem?" I snapped at him.

"What . . . who is that?"

"It's my mother."

"Is she all right?"

"No, she's not all right. She's dying."

"Is she in pain? Should we call somebody?"

"She's not dying right this minute. She just found out she's going to die and she's not too happy about it." I was surprised to hear how bitter I sounded. I was embarrassed and angry at my mother for making her pain public, for making such a scene. Why didn't she at least close the windows? And I hated this stranger, with his cut-off shorts and tanned legs, for intruding on our family tragedy. I snapped, "I'm sorry if she's bothering you."

"It's not bothering me, I just . . ."

Moana pulled my arm and got me into her car. "He's just concerned."

Of course he was concerned. That howl like an animal with its foot caught in a trap was disrupting his quiet suburban neighborhood. The thought of that sound coming from my ladylike mother, with her lipstick and pearl earrings and soft telephone voice, was almost as frightening as death itself.

Six months later, I was pregnant with Ambrose and had been on bed rest for weeks when one of my sisters called to tell me they were all ("except me, except me," I thought, over and over) gathered at the hospital. Despite my father's round-the-clock ministrations, my mom had been in so much pain at home that she had asked to go to the hospital. The doctors gave her a few days to live. I wanted to go immediately to her side. With Emlyn born almost six weeks premature, my doctor feared that this baby would be born even earlier. He refused to sign a waiver allowing me to fly.

My sister told me later that after she had gotten off the phone with me, my mom gestured and said my name with a sense of urgency. "Do you want her to come?" my sister asked. "NO!" my mom said in a hoarse whisper. My mother's refusal to see me was the ultimate proof of her love. I knew she wanted to see me, to squeeze my hand, and have me kiss her cheek one last time, and above all, to have the security and satisfaction she had always gained from having all her children around her. There was no doubt of that. But at all costs, she wanted to spare me the kind of loss that had haunted her for her entire adult life. She had lost two babies during the years my father served in the navy. I have a picture of her from those sad days. She is sitting in a corner of the tiny skid shack they lived in outside Kodiak, Alaska, wrapped in a silk dressing gown my father brought back from Japan. She looks heartbreakingly beautiful, her face ash white, her eyes great dark drowning pools.

All my life, in certain quiet moments, I had heard my mother make a kind of reverse sigh. She would take in a sudden breath and it would catch at the bottom of her lungs for a moment before she let it out. It was a deeply sad sound, like a whale dying on the beach. I remember it especially on winter afternoons on the way to pick up my sisters from their ballet lessons. As we waited for a green light at a busy intersection, I would listen to the tick of the turn signal, the slap of the windshield wipers, and that shuddering sigh. When I was looking for ways to deal with Emlyn's colic, I read about homeopathy. The book of symptoms listed a condition characterized by deep and repeated sighing and attributed it to long-held grief. If only it were so simple, I thought. A small white tablet placed on my mother's tongue, and all her old griefs would melt away.

Three days after my mom entered the hospital, I had to say good-bye to her over the phone. Emlyn lisped his newest phrase, "I love you," and I talked fast, trying to fit in all I wanted to say. When her breath grew ragged and she said faintly, "I have to go," I wanted to protest, "No, I'm not ready. Give me a few more minutes. Give me a few more weeks. Just wait for me, Mom. Wait, Mom—" One of my older sisters came on the line and said she was too weak to talk any more. That evening my mother stopped talking altogether and in less than twenty-four hours she was dead.

She died on a warm summer day in June. It was late afternoon—her favorite time of day, when, her household chores finished, she would remove her apron and take the seven of us down to the beach. She would settle into the sand, digging in her heels and slipping the straps off her freckled shoulders with a groan of satisfaction. Occasionally she would rise and, shading her eyes against the setting sun, scan the waves, counting heads in the water. Sometimes my father would join us when he got home from work. Wrapped in a towel in the cold sand, I watched my parents swim together. I saw my mother, then, just as my father remembered her as a teenager—emerging from the water, laughing, droplets of water clinging to her black curls.

I wish that I had been able to do one final thing for my mother—smooth back the hair on her forehead or feed her a spoonful of ice cream. I wish I had been there the night she called all my sisters and brothers together for the last time. They stood in a circle around the bed holding hands while my mother gave them her blessing.

Ambrose came and sat next to me. Shivering in the shade of the trees, he showed me some pretty rocks he had found in the creek. As Bertie and I talked, my son sat in quiet absorption, examining each rock carefully before putting it in his pocket. Ambrose was born six weeks after my mom died. I had my first labor pains on what would have been her fifty-fourth birthday. Ambrose was born at noon the next day. Three weeks premature, he was a small, quiet baby, with a cry so soft that I had to keep a sharp ear out to hear when he awoke from a nap.

I took to carrying Ambrose everywhere. I joined a mom and baby group, and, as we sat in a circle on the floor of the community center, I looked around at all the cheerful, chubby-cheeked babies. They played with toys, crawled, explored, and climbed. Ambrose sat silently watching everything with a slight frown crumpling his small brow. He was so serious that the other mothers called him "the little professor." I often wondered how he had been affected by those motionless weeks in bed listening to the sound of my crying.

We left the cool spring and hiked back down the canyon. It was easy going downhill, and Ambrose skipped ahead with Bertie while Liz and I followed with the horses. Liz broke off a few stalks of miner's lettuce and we chewed them as we walked. She talked about the various uses for plants—which ones are good to eat, which ones are poisonous to cattle. Camas root is good for colds; sage soothes a sore throat. I thought of an old remedy I had read about in a book of Idaho folklore. The settlers had believed that cancer could be cured by applying a compress of cobwebs. I imagined myself going through the house on Knob Hill, Uncle Herman's farmhouse, the skid shack in Alaska, going back farther and farther to gather enough cobwebs to staunch my mother's pain.

Liz and I were still concerned about Bertie, but it later turned out, as it often does, that we were worrying about the wrong thing entirely. The biopsy would turn out to be negative, yet the long-awaited visit from her parents never materialized.

I heard Ambrose shouting and singing before he rounded the curve in front of us. He ran toward us yelling, "I saw a bear! A bear!" Bertie followed him back up the hill, grinning. "We came around a bend and there was a yearling cub. He ran across the creek and then just turned around and sat on the hill looking at us for a few minutes," she said. Ambrose turned and marched ahead of us, chanting his refrain all the way down the hill.

Bertie and Ambrose arrived back at the house at Burns Ferry first. It was getting late, and Bertie wanted to get dinner started and check the answering machine. Her mom was supposed to leave a message about plane reservations for their trip. Liz and I came through the dark alder thicket and led the horses along the road lined with apricot trees. I could smell the overripe fruit split open in the grass: the remains of the June drop. Dimly, we saw Bertie walking toward us across the pasture. When she got within hailing range, she raised one arm like a dream figure in the dusk and called out in a distant voice, "My father died today."

TEN

Brief Lives

Like other suburban children of the 1960s raised on Lassie and Flipper, my siblings and I had an unrealistic view of animals, but it was an oddly hopeful view: it felt good to believe that animals were our loyal friends who would go to great lengths to save us. But for our children's generation, the tables have flipped, and it is they who are expected to save the animals. They've grown up with images of burning rain forests, oil slicks spreading from wrecked tankers, and deformed frogs.

My son Emlyn, overwhelmed by the uniformly bleak messages he received from the media, became convinced that all was already lost. "There won't be any environment left by the time I'm grown up," he said at about age six. He turned his attention to the creatures and aliens in fantasy novels and computer games.

His brother, Ambrose, reacted differently. He read his *Ranger Rick* and *Zoobook* magazines, joined an organization to save the lions, and pored over grisly pictures in *National Geographic* of rhinos killed for their horns. It was all so distant and focused on loss that I was glad he had some positive experiences to balance things out. He spent many afternoons in a field at the edge of town, just a block from our house. There he caught frogs, praying mantises, garter snakes, and grasshoppers and followed rabbit and deer tracks in the snow. He discovered a clutch of pheasant eggs and begged me to let him hatch them in an incubator.

A small logbook he got as a Christmas present got Ambrose thinking about a different way to "keep" the animals he saw. He stopped snatching at anything that moved and began sitting still to watch instead. He used the logbook to keep track of the animals and birds he saw each day, whether it was a muskrat at the city park or a buck gazing at him from the brush along the Snake River. Instead of live animals, Ambrose started bringing home animal parts—a cat skull, a muskrat foot, a pheasant's wing, miscellaneous bones. He seemed particularly fascinated with teeth; a jawbone was a good find. He handled these things not with scientific curiosity, but with a kind of respectful attention. He would carry his latest find around for days like a totemic object, setting it on his bedside table at night.

One of our homeschooling friends used to pick up roadkill and store it in his freezer until he had time to show his nine-year-old daughter how to stuff and mount the animal. We too started watching for roadkill. Despite Ambrose's long-standing desire for a quill, neither of us could bring ourselves to touch a dead porcupine by the side of the road. Fascinated by the sight of the split-open belly, the bluish organs spilled out onto the pavement, he asked, "Is that the heart? Are those the lungs?" We found a gray form at the foot of a power pole that turned out to be a great horned owl. Electrocuted, it must have dropped like a feathered stone. Walking along a county road, I turned over a pile of feathers; it was a kestrel struck by a car in mid-scream, its beak still open and angry. Its feet were wonders of functionality: hooklike talons and soles as rough as sharkskin to grip its prey. One winter day we found a possum curled on its side, frozen to the ground. Ambrose squatted down and studied the whiskers outlined with frost, the pink toes. "It's so . . . perfect," he said.

Sometimes, Ambrose learned, the reality of live animals is not so perfect. A Beanie Baby bat, wrapped in its own soft wings, is one thing; a creature zooming around your head when you turn on the bathroom light is another thing entirely. On summer nights the house on the Burns Ranch is left wide open to get a cross-draft; bats swoop through the house chasing moths. Liz will sometimes lose her patience when a moth circling the reading lamp is nabbed just inches from her head. She'll knock the offending bat out of the air with a broom, throw a towel over it, and release it outside.

We often spent Ambrose's birthday at the ranch because it coincides with the Perseid meteor shower; the dark canyon sky is perfect for watching falling stars. Ambrose remembers his eighth birthday primarily because of the bat in the bathroom. I remember that night for the rubbery flapping of wings uncomfortably close to my ears; despite the heat, I slept with the sheet over my head.

Liz's tolerance extends toward snakes as well. Unlike many westerners, who kill any rattlesnake they see, the Burnses leave rattlers alone unless they come into the yard. Relaxing one day with the door open to catch the breeze, Dooley heard a rattler at the foot of his bed; he hurled a boot at its head and went back to his nap. Another time I went out to the orchard just long enough to pick a bucket of peaches. When I came back, Bertie was standing at the kitchen sink cleaning a rattlesnake skin. In the few minutes

I had been gone, she had walked out the door to move the sprinkler and almost stepped on the snake in the grass. She chopped off its head with a hoe, stripped the skin, and threw the carcass to the dogs. She would tack the snakeskin to a board to dry and then hang it on the wall along with the other skins and furs.

In over ten years of going to the ranch, I've seen a rattlesnake only once. Discovered by a tour guide on the path at Ram's Head Rock, it retreated under the old tractor by the outhouse. Tourists lay on their stomachs—a few feet from its head—to get a flash photo of it. I felt sorry for the thing and later described it to Liz. "Sounds like a pygmy rattler. I haven't seen one of those in a long time," she said. Liz says she's noticed that, in general, there are not as many "little things"—frogs, snakes, butterflies, bugs—as there were when she was a child. But when the river recedes in late spring, I've seen yellow-and-black butterflies rise by the hundreds from the puddles left behind. And there seemed to me to be plenty of bats in the canyon. One time Liz took me to the top of a rocky pinnacle by the river to show me where the bats slept. Shielding our eyes against the sun, we peered into a deep crack in the rock until the lumps against the wall turned into small folded bats, stirring a bit like babies in their sleep.

Curious to find out what "little things" in particular have been disappearing from Craig Mountain, I obtained a wildlife inventory recently completed by the Idaho Fish and Game Department. They have documented the decline of many small creatures in the area, from Merriam's shrew to the tiny western pipistrelle, a bat that lives in cracks in the rocks such as the one I explored with Liz. Many of the smaller birds are disappearing too: yellow warblers, winter wrens, the brown creeper that goes a tree trunk in a spiral, and the pygmy nuthatch that goes down the trunk headfirst.

Although there are fewer frogs than Liz remembers as a child, they are still doing better on Craig Mountain than in many parts of the Northwest. Frogs are disappearing from ponds and streams worldwide for reasons that are not yet clear. Both spotted frogs and western toads are declining or have disappeared entirely from parts of Oregon and Washington, crowded out by non-native bullfrogs and eaten by introduced species of trout and warm-water fish.

One recent winter, Liz had some friends pull up and burn the tangle of blackberry bushes in the old trout pond. Liz dug out the overgrown, silted-in channel so that water again flowed into the hollow. With the trickle of creek water and the spring rains, the pond slowly filled. Out of nowhere, it seemed, scores of frogs appeared to lay masses of foamy egg bundles along the water's edge. When Ambrose went to stay that summer, he walked around the pond with Liz, watching the tadpoles swirling like animated musical notes just beneath the murky surface. By the time I came to visit a few weeks later, the blackberries that Ambrose and Bertie had gathered by the bucketful were gone and so were the frogs. "Where'd they go?" I asked Liz. "Hopped on up the mountain," she said. I was disappointed that the frogs had come and gone and I still had never seen one at the ranch. I asked what kind they were.

"Your basic frog with spots," she said.

The spotted frogs on Craig Mountain are reproducing well because they live in ponds and small puddles, well away from the warm-water fish and bullfrogs that have been introduced into the Snake. The status of the tailed frogs is more precarious. They live in fast-moving streams up on the mountain, but there are only a few isolated pockets of them left; any of these could be wiped out by even a small logging operation that slowed and warmed the streams by filling them with silt and eliminating shade cover.

Intensive logging on the mountain, particularly over the last ten years, has eliminated much of the old-growth ponderosa pines. Without the ponderosas, cavity-nesting birds such as the nuthatch are finding it harder to survive. The piping note of nuthatches can often be heard in forests throughout the Northwest, but other cavity-nesters are seldom seen or heard. One that is disappearing from Craig Mountain is the flammulated owl, an owl the size of a bluebird that migrates from Central America each summer. Another is the northern pygmy owl, which I've never seen, but which I recognize from an entry in Smokey's Alaska journal. She wrote of being spooked by a "little man" looking in the window of the cabin. It was a pygmy owl using the windowsill as a perch to survey the beach; the dark feathered patches on the back of its head looked like eyes peering in the window.

The connection between logging and the decline in cavity-nesters on Craig Mountain is fairly easy to imagine, if not prove—each ponderosa felled is one less tree in which a bird can build its nest. The effect of cattle ranching on songbirds is more subtle and harder to assess. Cowbirds, which colonize other birds' nests, tend to increase with the number of cattle. And many songbirds build their cup-shaped nests in shrubs along streams and in brushy draws—just the kind of places cows like to congregate. There are fewer yellow warblers than there used to be on the mountain, but no studies have been done to find out if cowbirds or cattle have anything to do with it.

There are fifteen species on the mountain that are considered "special status" because they are rare, endangered, or their population is dropping in other parts of their range. These include northern goshawks, ringneck snakes, and the white-headed woodpecker. But even with small numbers of these rare animals, there is no shortage of wildlife on Craig Mountain. There are eight kinds of snakes on the mountain, ten kinds of bats, nine kinds of owls, eight kinds of woodpeckers, six kinds of frogs, and seventy-five different songbirds. There are also various fur-bearers: river otters, beavers, mink, badgers, bobcats, weasels, skunks, porcupines, raccoons. Martens may be completely gone at this point. There are plenty of snowshoe hares but no longer any white-tailed jackrabbits. And Liz remembers possums, which aren't even listed on the current inventory.

Every spring and fall, a parade of pickups returns to town bearing elk, deer, turkeys, grouse, chukar partridges, and other game; but the flow of animals also goes the other way. People from town consider the countryside a dumping ground for unwanted pets. Like most rural residents, the Burnses have taken in many such abandoned animals, including whole litters of kittens dumped by the road. Friends also ask them to take their pets: a dog who has run away once too often, a horse they have no time to ride. Recent acquisitions include a turkey with a bad limp and a Vietnamese potbellied pig with a belly as huge and hairy as an old sailor's.

One pet we had a hand in bringing to the ranch was a particularly vicious ferret named Randy. My husband's friend Lucy, an opera singer, came to visit from California. Originally from Mississippi, she still speaks with a drawl, although with her hyperkinetic personality she sounds like a hound dog on speed. Lucy has an unerring instinct for making bad choices—in men, cars, even pets. What makes this foible endurable, even endearing, is that she cheerfully admits she has bad judgment yet seems genuinely surprised every time things turn out wrong.

Lucy arrived at the ranch carrying a cat travel cage held nervously away from her body. Bertie scooped up the cage and lifted it high, peering closely at the animal's weasely face. "I've always wanted a ferret," she said. Liz barely glanced at the animal looping back and forth. Instead, she took one look at ninety-eight-pound Lucy in a leather jacket and leggings, staggering up the beach in high-heeled boots, and muttered, "That gal needs to eat."

Lucy had decided to bring a ferret home as a present to her boyfriend. At a pet store in Lewiston—a rank-smelling place with a floor covered with sawdust, feathers, and rodent droppings—she had picked out a baby ferret, brown, with blue eyes. Problem number one: ferrets are illegal in California, so she decided to smuggle him onto the airplane inside her coat. She abandoned this plan quickly enough when she discovered problem number two: Randy bit. He not only bit, he hung on until he was pried loose. I had visions of Lucy on the airplane suddenly letting loose with one of those high notes as Randy sunk his teeth into her breast, practicing a little nipple-piercing for his new life in California. We called Bertie, who has a soft spot for all baby animals, and asked her if she wanted a ferret.

Bertie figured the ferret hadn't been handled enough, so she tried to get it used to human touch. It kept biting. The day it sank its teeth into her thumb and she had to whack it against the bedpost to dislodge it, she said, "That's it. I give up." She set it loose in the yard, and for the next couple of months it was seen running hop-gaited as an otter though the bushes. Occasionally it ventured onto the front porch to catch mice attracted by the feed grain stored in large bins.

Then one day, Bertie found the ferret caught in a pack rat trap in Dooley's workshop. He was dehydrated and too weak to even think of biting when she picked him up. For days he lay in her lap, a limp, long rag of a thing, letting Bertie feed him by hand. By the time he was well, he was completely tame and never bit again.

Like most of the animals on the ranch, with the exception of the horses and dogs, the ferret lived a rather brief life. (He disappeared a few months after his run-in with the rat trap.) Steers and dry cows were trucked to the auction house without regrets, although Bertie always moped around for a few days after sending off one of the calves she had bottle-fed. Rabbits were eaten by rattlesnakes, chickens by coyotes, cats by owls. The parakeets that died of the cold one winter were replaced by a pair of doves. Bertie mourned the mouse, a hand-me-down from Ambrose, that got a leg caught in its bedding and died of gangrene. Dooley and Liz have found fawns tangled in barbed wire and at least one calf mauled by a bear. The small herd of sheep was decimated by coyotes until all that was left was an ornery old ram that stalked around giving everybody the evil eye, trailing loose strands of wool. The cow dogs got after him one day and chased him far up the canyon; later one of the dogs returned carrying a leg of mutton with the wool still on.

Liz has told me about all the pets she had as a girl: raccoons, skunks, ground hogs, all kinds of frogs, lizards, birds.

"We even has a cougar cub for a while," she said, "until my folks sent it to a zoo. My mom raised an eagle, too. She used to catch fish for it and feed it by hand. Somebody was always finding an orphaned animal and bringing it home. Of course, you can't do that anymore. You're supposed to just let them die."

One of the few long-lived animals on a ranch is the horse. Horses are an exception to the rule that every animal on the ranch has to earn its keep, and many an old mare will continue to consume hay long after she's outlived her usefulness. Horses are a combination of working animal, wild creature, and pet. I'm sure many horse owners would add "friend" to this list. Alone of the animals on the ranch, horses flow effortlessly back and forth between the wild and the tame, between the inner and outer worlds of the canyon. In the far pastures deer, bighorn sheep, and elk move among horses unafraid. Closer to home, cows pay them no mind and dogs trot at their heels. They seem to be everywhere and nowhere. At night the sound of the horses clipping grass in the yard drifts through the bedroom window and enters my dreams; but in the morning when there is work to be done, the horses are nowhere to be found—only a curving line of horseshoe prints in the sand along the river.

Most farms and ranches in the Snake River Canyon keep just a handful of horses now. The days are long gone when every farm or ranch had a blacksmith shop. Hiram's old blacksmith shop next to the barn has stood unused for many years; the dozens of worn horseshoes on the fence rail beneath the window testify to the effort involved in keeping horses shod in a land as rocky as this. Now the Burnses' horses are shoed by an itinerant blacksmith who makes a yearly circuit through the intermountain West, visiting isolated ranches and farms to ply his anachronistic trade.

Browsing the classifieds in the local paper, I often come across want ads that would not have been out of place in a newspaper of a hundred years ago. In these days of high-tech careers, it's good to know that someone is still looking for egg pickers, swine handlers, and ranch hands. The farmer or rancher placing the ad will specify that they want a hard worker who is used to barnyard smells and what is tactfully called "animal debris." Some will come right out and state, "No cowboys need apply." Dooley remembers that back in the 1930s the largest landowner in the canyon refused to hire "genuine cowboys" because they were always so busy fiddling with their gloves or chasing after their hats that they never actually got around to working. Dooley has a different reason for looking down on cowboys: their close association with horses. As far as he is concerned, a horse is an expensive, near-useless creature that eats grass and hay that could be going to fatten up a cow.

Dooley couldn't understand it when Liz, like many girls raised on ranches, turned out to be horse crazy. Dooley said, "From the first time she saw a play horse or any kind of horse, she was just crazy about the sonofabitches. She always had horses around. One of them had a goddamn colt and he was down there on the sandbar. The colt got after another mare there and it kicked at her. Hit Lizzie in the mouth and broke a tooth off.

They had to put a bridge in there. She got to riding quite a bit and another goddamned silly-shittin' horse run off with her and run into a tree and damn near tore her leg off. She still got a stiff knee from it."

Riders who have spent some time on horseback in the canyon have learned, often the hard way, to take their feet out of the stirrups on a steep slope, just in case the horse slips. As Liz said, describing a recent leap from the saddle, "When they start to fall, I say good-bye." Once when Bertie was following a cow up onto the rimrocks, her horse lost its footing and fell over backward. Bertie said, "I saw her falling towards me in slow motion. She landed on my legs and rolled right up to my neck. I could feel all my organs squishing from bottom to top." Bertie was bruised and sore, but grateful something hadn't ruptured. When I saw her weeks later, she was still walking bent and slow like the rodeo bull riders who creak around the corrals at the Lewiston Round-Up.

It's a mystery to me why people like Bertie and Liz still love horses even after being injured by them time and again. Dooley explains it like this: "Horses do somethin' to people. Just wipes out their brains. That's why these cowboys go out there and try to kill themselves. They get close to a horse, they just lose every iota of brains they ever had."

Liz usually hires a cowboy named Dermott, a horse breaker from British Columbia, for several weeks a year. Bertie doesn't like how rough he is on the horses and complains, "I don't know why you hire that guy." Liz's reply: "He doesn't drink, and he's available for long stretches at a time. There aren't too many cowboys you can say that about."

Without a bunch of hollering cowboys in ten-gallon hats to move the cows around, Liz has to rely on a totally different approach to working with cattle. She works up close to the cows, often on foot, and she tries to move them slowly and easily, without getting them stirred up. I had heard of other women ranchers who have developed this same technique of moving the cattle by intuition, good timing, and patience. Cows that are stressed eat less, produce less milk, and gain less weight. By moving the cattle with a minimum of fuss, Liz is actually maximizing her profits. This method works most of the time, but even Liz can be caught off guard sometimes.

One spring Liz was working alone down at Ram's Head; she was concerned to see a newborn calf lying by itself, with the mother standing a fair distance away, ignoring it. Keeping an eye on the mother, she had started to lift the calf to a standing position when she felt a blow from behind. Another cow had charged up and hooked Liz beneath the arm. The angry cow backed off and Liz kept on working, ignoring the growing stiffness in her arm. When she got back to the house, she peeled off several layers of winter clothing and discovered they were soaked with blood. The horn had gouged a four-inch-long gash in the underside of her arm. When Liz showed me the scar weeks later, it was still red and sore-looking. I asked if they had gotten rid of the cow, and Liz said indignantly, "No, but I was mad as hell at her. I'd have understood if it had been the mama cow, but that wasn't even her calf!"

Dooley, who disapproves of his daughter's method of working with cattle, said, "Lizzie never had sense enough to be cautious. She's still that way. She's too damn brave around 'em. To wrangle a cow, you gotta baffle 'em, you know, spook 'em away from you or whatnot. 'Cause if you turn around and run they'll knock you down. Maybe you can shoo 'em back and maybe you can't. Just depends on how much power and guts you got."

I have no doubts about the amount of power and guts Liz has. I've seen it demonstrated on many occasions, one of which was when we were separating calves to be weaned before winter set in. One cow—not a heifer, but a large, full-grown cow—became enraged when her calf was herded into another corral. The cow lunged this way and that, glaring at me behind the gate and at Bertie against the barn. She seized on Liz, smack in the middle of the corral, as the most likely target. In the ankle-deep slop, there was no way Liz would be able to scramble out of the way if the cow charged. (That day, both Liz and Bertie had fallen full length in the mud. I managed to stay on my feet because I was working the gates and so had something to hang on to.) The cow charged. Liz shouted, "Get back!" and the animal skidded to a stop less than six feet from her. Liz stood her ground, staring the cow in the eye. It was strangely thrilling to be hovering in that moment when things could go either way: either the cow would back off or Liz would be trampled, perhaps gored.

Just then Ambrose bounded up to the fence, calling, "Mom, can I borrow your camera? There's a bald eagle down by the river." Liz didn't even look at Ambrose, but I could see her attention waver. Instantly the cow stepped forward and shook her long, splayed horns. Liz said, "Ambrose, don't distract me!" Her focus narrowed down to a point like the tip of a rapier held within inches of an opponent's breastbone. For several long moments, the cow kept her head lowered, her tail flicking. Then she wheeled and trotted away.

Liz gave her characteristic snort when I told her that Dooley thought she took too many chances around the cows. "That's just because he hates cows. The only thing he hates more than cows is horses. His idea of herding cows is to ram them with the pickup."

Bertie backed her up. "Yeah, one time I saw him go after a cow that was getting away. He chased her with the pickup and got in front of her and turned so that she ran right into the truck. It worked. She stopped and went back the way she was supposed to."

"Sure it worked," Liz said, "But when you handle them like that, they end up aborting."

Dooley summed up his philosophy on moving cattle with the matter-of-fact statement: "You gotta draw blood on 'em to get 'em anyplace." He has no patience for cattle that put up more than an ordinary resistance to being herded. Every few years, they end up with a bull that just won't stay fenced. One giant refused to be herded back up the mountain, so they tried to move him in the pickup. They got him cornered and into the truck, but he jumped over the rack, smashing the window on the driver's side and bending the top bar of the rack.

"To hell with him," Dooley said. "Send him to auction with the next load of dry cows. Let someone else deal with him." The next time a bull became intractable, they resorted to a simpler solution. One evening while Liz was cooking dinner, I noticed a new Polaroid pinned to the wall. Dooley, relaxed and smiling, had his arm slung casually around the neck of Big Red, the biggest, orneriest animal on the ranch. I was about to ask how in the world he had gotten that close to Big Red, when I saw a bullet hole over the bull's eye. It was just the head mounted on a fence post. Liz laughed and gestured at the hamburger in the pan and said, "Yup, here's what's left of Big Red. We've got a whole freezer full of him."

In the middle of dinner, Ambrose paused to look at the rather gruesome pictures of the bull being butchered in Dooley's workshop. It was enough to turn my stomach, but Ambrose just looked from the photos to the meat on his plate and said, "I'm eating Big Red? Cool!" Then he sat there, chewing thoughtfully.

Ambrose has watched and absorbed how animals are treated at the ranch, whether it's the congenial greeting Liz offers to a coyote suddenly appearing over a rise or Dooley roundly cursing the horses for nosing in under the trees when he's trying to pick apricots. Taking his cue from Liz, Ambrose has adopted an attitude toward animals that is both respectful and practical. I am grateful that he has managed to avoid the extremes of both the city kid's and the country kid's views, realizing that animals are less than the idealized image depicted in nature magazines and yet more than simply objects for target practice.

Although Dooley has raised cattle for over forty years, he looks back fondly on his shepherding days. "I loved the sheep. Sheep is a more calm-natured animal. They're not so belligerent as cattle. Cattle, they're up there on the mountain and the only time you see them is when you round them up or move them or something. Whereas the sheep, you're with them every day. Kind of like the difference between an investment and running a factory."

Liz insists that her father has always liked working with machinery more than with animals, and that for him ranching—sheep or cattle—has always been just a way to turn grass into cash. But I think even Dooley has an appreciation for animals and that near-miraculous process by which they come into this world. In his letters to me, he sometimes refers to the horse and cattle as "live stalk." In the spring, when the grass turns a tender green and the calves emerge, each one as simple and perfect as a radish pulled from the dirt, I can believe they are live stalk, growing right up out of the earth.

ELEVEN

Blue Heron

Liz shouted above the roar of the truck, "You sure you want to stay, Weezer?"

"It's just overnight. I'll be fine."

She nodded and tossed me the key hanging from the rearview mirror.

"Don't lose it. Lock up if you go anywhere."

The truck made a wide U-turn in the dry grass, whacking down tall stalks of thistle. Bertie sat on a toolbox in the back of the truck. She grabbed for a handhold as they bounced over a rock, then waved good-bye.

At the barn, they turned down the road that followed the Snake River to the other end of the ranch. I could hear the truck for a long time, the sound fading in and out as it went down one ridge and up the next. The muffler on the truck had fallen off again, and Liz said she was tired of wiring it back on. In the dust on the dashboard Liz had recently drawn one of the petroglyph figures from Ram's Head Rock, with the signature triangular body and upraised arms. In one hand it held a hammer, in the other a wrench.

I stood and looked around, uncertain what to do with myself. Up on the rock, the new windows in the cabin glinted in the late afternoon sunlight. I was grimy and tired after working on the cabin all day, and I thought longingly of the claw foot bathtub at the Burns house. But I would be the first person in over twenty years to spend the night in the cabin, and that was worth more than a hot bath.

I walked down to the river to wash up, crossing the pasture and taking the path that led around the base of the sixty-foot-high rock. Where Liz's childhood home used to stand, tucked against the east side of the rock, there was nothing but a charred spot of ground. I almost regretted what a thorough job we had done of demolishing and burning the old shack the previous winter. I kicked at the pile of ashes and turned up only a broken bit of china. Last winter, I had found an unbroken green willow pattern plate and set it aside. Liz glanced at it and said, "I just threw a cup like that in the fire." I raked the hot ashes, looking for the cup, and took home the mismatched pair—the perfect plate, the cup discolored by the heat.

With the remains of the old house burned, the junked cars hauled away, and the new roof put on the barn, Ram's Head was starting to look less like an abandoned homestead and more like part of a working ranch. It had been twenty years since Liz and her parents had moved to the house at Burns Ferry, two miles upriver. In the intervening years, several of the outbuildings had collapsed. The cabin's windows had been shot out, replaced, and shot out again. The cleanup effort had taken Liz several years, working in spurts whenever a few days could be spared from the regular chores, but now that the cabin was fixed up, perhaps she could breathe a little easier. With the flag flying overhead and a couple of bright rag rugs flapping on the clothesline, the place looked lived in. Perhaps boaters on the river would now think twice about shooting out the windows or throwing their garbage on the beach.

I went around the north end of the rock, past the doorless outhouse and the tractor overgrown with grass, down to the white sand beach. I slipped out of my dusty clothes and swam in the cool water, being careful to stay where I could touch bottom. Just a few feet from shore, the current restlessly coiled and uncoiled itself into whirlpools. Here, along the deepest stretch of the Snake River, the eddies are strong and unpredictable. Even Liz wears a life jacket when she swims alone.

Before Hells Canyon Dam was built in 1968, whirlpools strong enough to swallow trees used to form at Ram's Head. Every spring the high mountain rivers—the Salmon, the Grande Ronde, the Imnaha—washed dead trees into the Snake River. Huge cedars and pines, forty or fifty feet long, would get pulled into an eddy, sometimes thirty feet across, with a vortex ten or fifteen feet deep. As they rounded the bend at Ram's Head, the heavy root ends would get sucked down first, the crown circling clockwise for a few minutes. Then the whole tree would be yanked upright just before it disappeared into the hole with what Dooley described as "a loud slurping noise."

Liz, who as a child often sat on Ram's Head Rock and watched the eddies, would hold her breath, and within minutes, a roaring boil of water would erupt, spitting out tree branches, splintered wood, and the bare trunk of the tree.

The dams have slowed and warmed the river. When the water is high, you can still sit on the rock and watch the eddies form. Although not as powerful as they once were, they are still strong enough to flip a raft or suck down a canoe. One recent spring, a canoe overturned near the rock and spilled three young men into the cold water. The men clung to the canoe until it drifted into a large eddy. One man struck out for shore, another got out into the main current and let himself float, but the third was caught in the whirlpool. A family in a jet boat came along and threw a rope to the man struggling in the eddy just before the water closed over his head.

Layered between sections of bare dirt, shale, and bunchgrass, rimrocks climb like receding stair steps from the river. Rock slides on the sheer slopes often send down showers of dirt clods that burst on impact on the road below. Drivers coming around the curves on the gravel road have to watch for rocks the size of motorcycle helmets. Gradually the top of the canyon has widened, until in some places it is ten miles from rim to rim. But the bottom of the wedge-shaped canyon is still narrow and winding. The clean, clearly marked west rim, where the sky turns a corner onto the Anatone Prairie, is two thousand feet above the river. Farther south, in the deepest part of Hells Canyon, the mountains rise seven thousand feet above the Snake. The gorge is deeper than the Grand Canyon, although it is many years younger.

In the geology building at the university in Moscow, maps painted on the stairwell walls illustrate the process of continental drift. The Salmon Mountains of Idaho that rise over the east side of Hells Canyon were formed as an island somewhere near the equator. Eventually, the island chain moved north to join the rest of the landmass of Pangaea. The maps show Idaho in red, a little broken cookie of a shape, drifting across a flat blue sea. Finally, as you climb the stairs, Idaho bumps into Pangaea like a small, homely spacecraft docking with the mother ship.

Ram's Head Rock is one of the few remnants from that earlier time when all the continents were one. A chunk of metamorphic rock formed two to three hundred million years ago, it is distinct in shape, form, and age from the younger basalt of the canyon walls, which were formed by a series of relatively recent volcanic eruptions between six and seventeen million years ago. Every few thousand years another layer of lava would ooze over the land, baking the layer of soil to a reddish color. More soil would eventually be laid down, grass would grow, and then another eruption would occur.

Up on the Palouse, we live on the surface; the farmers in their wheat fields around the small towns barely scratch the dirt. The bedrock is hidden beneath some of the deepest, richest topsoil in the country. (Almost twice as much wheat per acre can be grown than in the Midwest.) I've heard various claims about its depth, anywhere from fifteen feet to sixty feet deep. Here in the Snake River Canyon, the rock is exposed, and I find looking at the layers reassuring—a reminder that it's always been this way: just when life gets back to normal, another catastrophe occurs.

I climbed onto the immense rock that rises abruptly from the sandy beach. My feet, still numb from the cold water, stung every time I jumped from boulder to boulder. At the boat pullout on a small gravel beach, a high slick rock face reflected the afternoon sun. Shading my eyes, I could just barely see the figures I knew were there—the red people with snail antennae, holding aloft sticks shaped like barbells. I sat on a shelf jutting out over deep water and watched the fish leap out of the water and land again with a smack. An otter appeared, first its head in one place and then its tail in another, like a magician's assistant who's been sawn in half but can still wiggle her feet.

Eddies and undercurrents formed a mosaic like ice floes on the surface of the river. I watched the water for a long time, mesmerized by the shifting patterns. I lay down on the warm rock, drowsily imagining it moving, slowly like a great heavy ship, across an ancient sea. The breeze blowing from the water suddenly felt cool on my legs. I sat up and saw that the sun had slipped behind the western ridge. The jagged canyon rim was so sharply etched against the sky that I could have folded it like a pop-up card and slipped it into my breast pocket.

I climbed the cabin steps—boards sawed, planed, and hammered in place just that day. The new steps felt solid under my feet and I clomped heavily with my boots for the sheer pleasure of it. Startled by the sound, a kestrel returning to her nest in the north wall immediately flew out again. The knothole must have been quite small, for there was a great deal of knocking and banging from inside as the bird got herself turned around. She emerged looking flustered, like a woman stuck in a restroom stall who comes out smoothing her dress, hoping no one was watching.

Inside, the pungent odor of fresh-cut wood masked the smell of mildew, mice, and dust. I felt around in the cooler for something to eat. A couple of bags containing leftover sandwiches—blown up and sealed airtight—bobbed about like fish bladders in the cold water. There were three or four round things: plums. I bit into one and got out my flashlight to read the titles of the musty paperbacks that had been moved from the closet to a new shelf on the south wall. There were a few westerns and mysteries, and one or two movie star biographies.

Today Liz and Bertie had worked slowly but steadily in the heat. They had a small table saw powered by a portable generator set up on the porch.

I admired Bertie's competent use of the saw and the way Liz leaned into the plane, peeling away translucent curls of wood. I've had an aversion to power tools ever since I got my hair caught in a wire brush attachment on a drill when I was a teenager. It had taken only a fraction of a second to wind my waist-length hair all the way up to my scalp and over an hour for my mother to untangle it. So today I had stuck with simpler tools—hammers, screwdrivers. But after I hit myself in the head with the hammer while trying to nail a board into place above the bed, I gave up and turned to cleaning and running errands. I swept and scrubbed the floor. I ran down to get things out of the truck and refilled the water jug from the pipe down by the old homesite. But mostly I cleaned up trash around the cabin. Over the years, hundreds of people must have sat on this porch drinking beer and tossing their empties onto the rocks below. I carted several garbage bags full of bottles and cans down to the truck.

As a finishing touch, Liz had hung a framed print on the cabin's east wall. The original was an oil painting of Ram's Head done by a local artist. It was all there—the cabin, the hump of rock, the swirling river, the huge log that has been stuck on the rocks for years now. I half-expected to see a tiny figure in the doorway; in a moment it would move inside and a pinprick of light would appear in the cabin window.

I realized that I feel at home here because I am part of the picture. Unlike the wilderness, where people are considered intruders and every step scars the land, this place is bound up with human history. It seems to accept my presence. Sometimes when I try to sound out a place, I get nothing back but a dull thud. This sere brown land, however, responds with a hollow, satisfying note, like a perfectly baked loaf of bread rapped with a knuckle.

As the sun went down one evening, Bertie lifted two drums from the back of the pickup. She plopped down on the ground and started thumping out a rhythm, motioning me to join her. As usual, I would rather have just watched and listened, but I took the drum and followed her lead. The drums were some of the larger ones Bertie had made, and it took a pretty good whomp to get a sound out of them. With each beat, I felt the shock travel up my arm and into my head where it buzzed at the roof of my mouth. The rhythm bounced between the rocks and the canyon walls, doubling back upon itself and creating multilayered echoes. There's been human activity at Ram's Head for thousands of years: ceremonies, feasts, rituals, no doubt many of them involving drums. The land seemed to recognize and welcome the sound, responding like a cat arching its back beneath a human hand.

The slam of a car door brought me out onto the porch. Across the river, a car had parked along the road, and a man was walking down to the beach.

A tow truck pulled up and drove onto the sand, backing down to the water's edge. The driver got out and joined the other man. They stood with their hands in their pockets, staring at a spot in the river.

A sheriff's car arrived. One of the men climbed into scuba gear. He consulted with the sheriff, put on a pair of flippers, and walked backward into the river. After a few minutes he surfaced and shouted, his voice sounding clear across the water, "I don't know if the cable's gonna be long enough."

The tow truck backed down almost to the water's edge. I hoped they wouldn't get stuck. I had sat by the river for hours on summer nights, listening to guys curse as they leaned out the window on one elbow and watched their wheels spit sand. They usually gave up and flagged down another truck to pull them up the steep riverbank with a chain. The diver disappeared again, staying under for what seemed like a very long time.

Before the dams were put in along the Snake, huge ice jams used to form at the bend in the river at Ram's Head. The backed-up water, forced beneath the ice, scoured out the river floor to a depth of 140 feet. Since Hells Canyon Dam was built, the hole has partially filled with silt, but it is still the deepest place on the Snake River. It's a dangerous place to dive. The unusual depth and width of the river here—about the length of two football fields—forces the water quickly around the bend, creating underwater eddies that erupt unexpectedly from the depths, swallowing boats and swimmers. There are also sturgeon down there, big shadowy things as long as a diver's body. To my relief, the diver surfaced and gave thumbs up. The tow truck hauled in the cable, bringing up a bright red late-model pickup streaming water from its bed.

People driving too fast on this stretch often can't make the curve at Ram's Head and wind up in the river. Many times it's a whole truck full of teenagers. Whether or not they live usually depends on how much they've had to drink and how cold the river is. I had no idea if this driver—from the looks of the vehicle, I would say a young man—survived his plunge into the river.

By the time all the vehicles were up on the road again, it was already dark. I sat on the porch and watched: the diver's car, the sheriff's car, and the tow truck with the pickup behind it retreated slowly down the river road—three of the vehicles lit by headlights, the fourth just a dark bulk.

I unrolled my sleeping bag on the platform bed, built to replace the old cot that now stood folded up in the corner. A pack rat scrabbled in the walls. The cabin had been a one-car garage elsewhere on the ranch until Liz's father had hauled it up onto the south end of Ram's Head Rock and converted it into an art studio for Gina. I wondered if Dooley had gone through all that effort in the hopes that, with a place of her own to paint, his wife would be more content to stay at home.

I had once asked Dorrie McAlpine what Gina had been like. She said, "Liz and Dooley walk the same, they talk the same. Gina was something different. She wore long skirts all the time. If she wanted to go barefoot, she'd go barefoot. If she wanted to smoke someplace, she'd smoke. People didn't know what to make of her."

It's easy enough to imagine a free spirit like that in California, but in small-town Idaho in the 1960s, she must have stood out like a Christmas tree in July. Gina had come from an even smaller town than Lewiston. She had grown up in Elk City, which is a scattering of cabins and old trailers just about as far into the wilderness as you can get in the lower forty-eight. Gina did some traveling when she was young, got hooked on flying, and ended up on the East Coast flying transport planes for the army during World War II. Even after she returned to Idaho and got married, she couldn't seem to get flying out of her system.

Dooley told me once, "We could barely keep from starving to death. But Gina had all these friends that had money and airplanes. Hell, she might go to town to get groceries and then call up from Tijuana, Mexico, or Nanaimo, Canada, or some goddamned place." He sighed and added, "She just would not stay home and tend to raising a family. She'd be here for a week or two and gone for a week. She just lived that way. She wasn't cageable."

It was an attitude that Dooley found hard to understand. He clung to the canyon like a burr, seldom straying far from the ranch. He had seen Chicago as a young man, traveling by train to deliver several thousand sheep to the city's famed stockyards. Dooley had done whatever it took to survive in the canyon. In lean years, he had supplemented his ranch income by shearing sheep, running a sawmill, or piloting a riverboat.

Like her father, Liz has seldom been beyond the Rocky Mountain states. She, too, does whatever is needed to keep the ranch afloat. The lean years now outnumber the good years, and she has spent several seasons working as a cook for a resort in Hells Canyon. The well-heeled tourists, having been herded on and off the boat in the blazing sun, are often peckish and demanding. Her generally tolerant attitude toward tourists is often stretched to its limit.

The air in the cabin felt very close, and I started wishing I could feel the river breeze. I turned on the flashlight and looked around for the noisy pack rat. It poked its pug nose out of a crack in the wall near the ceiling, eyed me with large, mild eyes, and disappeared. I got up and unfolded the old cot and moved it out onto the porch. I lay on my back in my sleeping bag. There was no moon. The sky and the canyon walls were equally black; I could tell one from another only by the stars—where the stars ended, the mountains began. I fell asleep with the booming of the river against the rocks sounding like the distant waves of my childhood falling on a darkened beach.

There is nowhere else I feel as sheltered and safe as this canyon. The high rock walls close out the world, and I feel at peace here. Liz scoffed at this idea. "Just wait till a cow messes with you. Then it won't feel so safe." Her bedside table is like a well-stocked first aid kit, crowded with liniment, Ace bandages, tweezers, aspirin, hydrogen peroxide, and Bag Balm, an ointment used on cow udders but which, Liz swears, is the best thing for chapped hands. Besides the everyday aches and pains, the cracked ribs and sprained wrists, she has suffered many serious injuries from animals. But I wasn't talking about ornery cows, skittish horses, rattlesnakes, eddies and swift currents, or any of the thousand other dangers on the ranch. It was a sense of refuge, I said, a feeling of being buoyed and sustained by this place.

Liz hesitated a minute, then said, "It's a powerful place all right, but it's not just a magical, happy place. Power can be used either way—good or bad. I know plenty of kids raised in this canyon who turned out rotten." She showed me a class picture from her elementary school days in Asotin, pointing out which kids had grown up to be wife beaters or pedophiles, had committed suicide, or had fatal accidents involving booze, boats, or guns—sometimes all three.

You could blame it on poverty or isolation. Or the kind of ignorance that made country kids shoot at anything that moved. But I knew Liz was right. There is a dark side to the canyon. I feel it sometimes on the river, when I am suddenly overtaken by a fear of being sucked down into the depths. I feel it when I wander too close to the Nez Perce burial site above the house at Burns Ferry. On winter afternoons, when the shadows topple from the cliffs, I can feel a cold weight like the pressure of sunlight against my head. In summer, the canyon walls reflect both heat and light until the air is so bright it sound in my head with an ominous buzz.

Ambrose feels the dual nature of the canyon, too. Sometimes he is happy to wander around at Ram's Head by himself for hours. One time we were sitting on the rock doing nothing when he asked me, "What do you want to do now?"

"Just what I'm doing," I said, thinking how rarely that was true.

Ambrose replied, "Yeah, me too," and picked up a pebble to throw in the water.

One day he watched the slow lift-off of a heron, skimming the water with its trailing legs, and lines of poetry started running through his head. On the drive home, he wrote a poem called "Blue Heron" in his journal, the lines slanting down the page in the wobbly handwriting of an eight-year-old:

> Soaring on the horizon
> even with the sun, a heron.
> Dost thou fly at the height of night?
>
> Clouds of gray with wisdom flaking from thy wings—
> blue feathers for the priest, his hut is his temple—
> flying to an unknown world.
>
> But we've been there in our dreams from the heaven way,
> soaring by the fire in the sky.

My husband too, witnessed that heron in the late afternoon light, brushing the water lightly with its wing, and his response was to write a liquid, haunting tone poem for solo flute.

On other days, Ambrose got spooked by the eerie silence of Ram's Head and begged us to leave. "It's too quiet." On those days he drew cat-eyed Modigliani faces, as inscrutable as the figures etched on the rocks. Emlyn, who can find good in almost any situation, managed to see only the negative side of things here. Falling on a prickly pear cactus for the umpteenth time, he cried, "Why did God make these things?"

Hells Canyon has always been a dangerous place. Early settlers shot each other over water rights and mining claims. They fell off cliffs and were mangled by farm machinery. Their children drowned in the river and their cattle died by the hundreds during bad winters and droughts. Steamboats exploded and ferries capsized. Chinese gold miners were murdered and thrown in the river. The Nez Perce were forced off their land, and settlers stole their cattle and dug up the possessions they had buried for safekeeping. There are sudden rock falls and freakish windstorms. And accidents happen with unusual frequency along this stretch of river. Over the years, countless cars, like the red pickup, have jumped the curve and landed in the river. Jet skis collide, rafts overturn, cars ricochet off the canyon wall and tumble down the riverbank.

One time a woman reached over to roll down her car window just as she rounded the sharp curve across from Ram's Head Rock. She accidentally opened the door, fell out, and was dragged down the gravel road. Another spring, a man jumping off the rocks slipped and hit his head, fell into the river, and drowned. A year later, in the same spot—not far from Burns Ferry—a fisherman decided to cool off by taking a swim. Eyewitnesses say he dived, surfaced "with his eyes open," and then was swept into an eddy next to a large rock. His body was found a month later, twenty-four miles downriver. His teenage daughter comes every week or two to leave flowers against the small white cross at the spot where her father drowned.

There are reminders of death everywhere on the ranch: a calf left beside the gate, its tongue protruding, the hide peeled back like a banana skin; the whiff of rotting meat; whitened bones in the grass. I find a curious comfort in being in a place where death is a daily occurrence. Among the cheerful, friendly farmers and the gentle landscape up on the prairie where I live, a long-held grief seems excessive—morbid, even. But here in this harsh canyon, a preoccupation with death does not seem out of place.

At Farewell Bend, farther south along the Snake, the Oregon Trail turned away from the river and led up over the mountains. There, pioneers camped on the flat for several days to wash clothes, repair wagon wheels, and repack their belongings to lighten the load for the arduous trip over the Blue Mountains. During the five-year period in which six of my closest relatives died, the bend in the river at Ram's Head Rock had been such a resting place for me—a place where I could pull out all my old griefs and rearrange them, sorting and stuffing and taking the cramped ones out to air. Small deaths— dead dogs, failed relationships, lost opportunities—were jettisoned to make room for larger and more recent ones. Then I, too, stood on the banks of the Snake River and said my farewells.

In the morning I padlocked the cabin and took a walk along the river before the sun rose above the canyon rim. Along this free-flowing part of the Snake, every bend has what is called a "meander bar" built up from silt deposits. Where the river makes a wide curve at Ram's Head it has formed an unusually large bar, a flat expanse a half-mile or more in length, rare in this nearly vertical land. Long before there was a homestead here, there was a Nez Perce village. When the Burnses lived here, there was a garden and a hay field and people coming and going. Now the land is used only as winter pasture. Until the weather turns cool and the cows are brought down from the heights of Craig Mountain, the bar seems huge and empty. I enjoyed walking along, swinging my legs, instead of climbing up and down hills and clambering over rocks. I found a good saw, its blade just slightly rusted, in the tire tracks along the edge of the pasture. It must have bounced out of the back of the pickup. I tucked it into my belt and headed back to the rock. The first boatload of tourists would be arriving soon. Throughout the summer months, jet boats take tourists upriver to Hells Canyon. Many of the tours include a stop at Ram's Head Rock for a quick look at the petroglyphs.

When Liz is here, she always goes down to say a few words to the tour guide. She doesn't greet the tourists, and she turns their questions over to the guide. Not because she can't answer them, but because the same questions come up again and again, and Liz has grown tired of them, especially stupid ones like, "How long did it take you to carve these pictures?"

Next to Liz, the tourists always look unusually clean in their white slacks and boat shoes. They stare openly at her stained coveralls and Olive Oyl boots. Liz would just as soon ignore the tourists, but it's important to make her presence known. She wants people to realize that this is private land and that someone is keeping an eye on things here. At the same time, she feels indebted to all those outsiders—environmentalists, river runners, archaeologists—who were instrumental in halting the Asotin Dam.

I once asked Liz why, if she can't stand tourists, she lets them come ashore to look at the petroglyphs or picnic on the beach.

"If it weren't for the tourists," she said, "Ram's Head would be under water." She recognizes that the local ranchers would never have had enough clout to stop the dam, no matter how much they stood to lose.

This tour group stood at the base of a tall rock face that extended up from the water's edge. In the morning light, the foot-high figures on the rock were much easier to make out than they had been the previous afternoon, when the light had reflected off the desert varnish—a type of patina that forms on rocks in hot climates. Hundreds of caves and rock ledges along the Snake River contain drawings, but many of them are now underwater behind one dam or other. Most of the sites are pictographs—designs and figures simply painted on the rock. Only a few sites, such as Ram's Head Rock, include petroglyphs, made by the much more laborious process of pecking the figures and abstract designs into the rock. With over five hundred individual pictures, Ram's Head is generally regarded as one of the two or three most significant rock art sites in the Northwest.

I always welcomed the chance to learn more about the petroglyphs. I moved close enough to hear what this guide had to say. Every local rancher, tour guide, and archaeologist has a different interpretation of the pictures. The guide explained that the figures can be interpreted rather literally as counts of animals killed in hunts, and of records of feasts. But a more recent theory, he said, proposes that the pictures were carved by shamans as a ritual demonstration of their power. The shamans wearing headdresses— the Martian-like creatures with antennae—sent their messages to the spirit world with the help of Coyote, a fat, cocky little figure that often appears in the lower right-hand corner of a panel of drawings. The sticks the shamans hold in their hands were often called "barbells" by earlier archaeologists. Now they do not hesitate to refer to them as spirit wands.

Dooley takes all these theories with a grain of salt. He's heard every possible interpretation over the years, and he doesn't think we'll ever know what the drawings truly mean. Liz understands them the way an artist understands a painting or a healer understands where the pain is coming from. But the petroglyphs are a mystery to me—both a comfort and a source of frustration. They are a constant presence whenever I am there, as though the rock itself were looking over my shoulder. Yet I am frustrated that I cannot make sense of the images. There are dots and whorls, circles, unidentifiable creatures, and strange figures doing who knows what. I keep looking for a story, even as I resist my urge to question everything, to truly "get" things by pinning them down and naming them.

The people who were interested in seeing the rock art close up moved to the front of the group. Those at the back stood with their hands clasped loosely in front of them, the way men often stand next to their wives in church—a stance that says, "I'll have you know, I'm not taking this too seriously." They glanced at me and looked up the path, trying to figure out where I had appeared from. The guide caught my eye and raised his eyebrows questioningly. I pointed at the cabin. Without breaking his patter, he turned and saw the cot on the porch and gave a nod.

The guide led his group up the trail to the other side of the rock, reminding them again to watch out for rattlesnakes. They walked along bobbing their heads—a glance up at the rocks and a quick look around their feet. The guide stopped in front of another set of petroglyphs, this one on the flat side of a boulder. The figures are smaller here than those on the large panel above the river. A long zigzag separates a young mountain goat from a poison lizard. Here is life and death, good and evil, youth and old age. Two figures with feathered headdresses dance, knees high, around a fire. The carvings stop at a natural crack in the rock, above which the face of the rock is stripped away.

"About ten years ago, somebody pried this top part off with a crowbar and stole it," the guide told the tourists, who clucked and shook their heads. But he did not tell them the end of the story as Liz had told it to me. The man who stole the painted rock got only halfway across the river when an eddy swallowed his boat. The rock sank and he drowned.

The guide showed them another set of petroglyphs that had been scratched and hacked at until the pictures are indecipherable. I wondered if any of these people had seen the petroglyphs on the rocks across the river. Some of the images were destroyed when the county blasted the rocks to widen the road. The petroglyphs are now so close to the road that you can throw a beer bottle out a car window and hit them—a thought which, judging from the amount of broken glass at the base of the rocks, must have occurred to a great many people over the years. Over half of the remaining pictures have been chipped away or disfigured by graffiti.

Liz feels a sense of urgency about protecting Ram's Head because vandalism has increased throughout the canyon. Dooley says that the petroglyphs were untouched until the 1960s. Few outsiders came up the river, and the locals respected the ancient sites. But in the thirty years since jet boats started taking tourists upriver, river traffic has increased 500 percent. The difference is most notable in the summer. Summer used to be a quiet time on the river. The water would be too low for steamers to bring supplies upriver or to take shipments of wool and wheat downriver. It was hot, and there was no reason to stay down in the canyon. The ranchers would move their stock to summer pasture in the mountains. They stayed cool in cabins beneath the pines and came down to the homesteads only to water their gardens and to can fruit from their orchards.

Now, despite the intense heat, summer is the busiest time on the river. The incredibly noisy jet boats, with their shallow drafts, can go far upriver even when the water is as low as twelve inches. Boaters from town flock to the river to fish, swim, and zoom around on jet skis. On a clear summer night you can drive along the river and see a string of campfires on the beach, one every mile or two, like torches lighting your way. The fires blaze just yards away from bone dry brush.

The tourists continued on up to the old homesite and stopped outside the fence that enclosed Gina's rock bowl. The guide explained that the rock had been formed millions of years ago when a lump of lava cooled into a ball and formed a hard shell; when the shell cracked away, it made two halves of a giant bowl.

Now that they were away from the cool breeze on the river, the tourists seemed stunned by the heat. The sound of trickling water coming from the cool bower of Virginia creeper drew them in. Liz didn't mind if tourists entered the fenced enclosure to take a drink; she built the fence to keep out cows, not people. The guide unlatched the gate, and they stepped forward, bowing their heads to avoid the low beam that has caught me on the forehead more than a few times. Some of them took a drink from the pipe. A couple of people threw coins into the twelve-inch-deep bowl. And one woman simply dipped her hands into the water.

After the tour boat left, I filled a tin can and watered the roses struggling to grow in the corners of the fenced enclosure around the rock bowl. On one of her rock-hunting trips before Liz was born, her mother had come across the basalt boulder near Cow Creek on the Washington side of the river. Dooley remembers that every time Gina went by the spot, she would get out of the car and go visit her rock. "She'd bend over it and make waving motions toward home." Seeing how much she wanted it, Dooley decided to bring the twelve-hundred-pound rock home for her birthday. At that time the ferry still operated at Burns Ferry, so he was able to wrestle the thing into his truck and take it across the river. Gina was delighted with the present; she planted flowering vines along a trellis and filled the four-foot-wide bowl with goldfish. Dooley hooked up a pipe to spray a fine mist over it all so that it stayed a cool, green bower even in the midst of summer.

Early one summer morning, I came with Liz to water the roses she had planted here. The roses were cuttings from her mother's rose bushes at Burns Ferry. Liz made the trip each day to water the small plants. She wiped her hands dry on her pants and stood looking out over the land. She said wistfully, "You should have seen this place. My mother had a huge flower garden. Everything was green. It was so beautiful . . ." But her mother is gone, and Ram's Head will never be the way Liz remembers it from her childhood. Back then, the shack was hidden by trumpet vines blooming with brilliant orange flowers. "You couldn't even see the house. It was like a cave in the rock." Liz bent down and plucked a handful of purple flowers. "Alfalfa," she said, chomping on a blossom. She handed one to me and I pulled off the head and ate it. It was sweet and grassy. Liz gestured toward the pasture, now mostly covered with cheatgrass and fireweed. "This used to be all purple as far as you could see. A field of alfalfa," she murmured, "thrumming with bees."

For many years, the rock bowl stood dry and empty. The flowers Gina planted had long since withered and died. After his wife's death, Dooley had trouble keeping up with the ranch, even after Liz moved back home to help out. He gradually turned the day-to-day responsibilities over to his daughter, although he often told her she was going about everything wrong. But even he couldn't argue with the idea of getting water to flow again down to Ram's Head; the risk of fire is an ever-present concern during the summer and fall months in the canyon.

Liz had a less practical motive for bringing water to Ram's Head; she was determined to make the land bloom again. But first she had to repair the pipe, which was clogged up or cracked somewhere along the line. She and Bertie bushwhacked up the overgrown trail to the source of the spring in a small cave. After taking a machete to the tangle of blackberry bushes covering the entrance, Liz climbed into the cave and stood waist deep in the cold water to clear away the accumulation of decayed leaves and debris that was blocking the intake screen. They watched the water flow freely through the screen and into the pipe and told each other that, with luck, maybe that's all that was wrong. They hiked back down the hill and across the pasture to the rock bowl, but there was no water coming out the end of the pipe. Not a drop. There had to be a leak somewhere. Liz puzzled for days about how to locate a break in a pipe that was several thousand feet long and buried a couple of feet below the surface.

One night at the community orchestra rehearsal in Lewiston, Bertie took me aside. Eyes round, voice a whisper, she told me that she had seen an amazing thing that day. "Liz found the leak."

"How'd she do that?" I asked. I imagined Liz digging randomly, her shovel leaving little piles of dirt dotting the field, like a cartoon figure trying to catch a mole.

"She dowsed for it."

"Dowsed?" I asked, not sure I heard right. "You mean with a forked stick?"

"Yeah, dowsing. But she used a pendulum, not a stick."

I associated pendulums with an old woman who lived a few blocks from me in Pullman. She put up large hand-painted signs in her yard warning that fluoride was poisoning our drinking water. She had once given me a rundown of her ideas on the subject. She had read every study ever done on fluoride, and I found her argument quite convincing until she got to the part about dowsing vegetables in the supermarket to see if they were safe to eat. She pulled her pendulum—a crystal on a fine chain—out of her purse and demonstrated how she held it over a bunch of broccoli in the produce section. "If it swings like this, it's safe. But if it goes back and forth . . ." She shook her head and giggled with the delight of a kid with a secret decoder ring.

My skepticism about dowsing faded as Bertie explained how Liz tied her mother's wedding ring to a thread and walked slowly over the pasture, going over the approximate course of the pipe. Bertie followed, carrying the shovel. "At this one spot, the ring jerked toward the ground so hard that the thread broke. It was incredible! I dug down to the pipe and it was all busted up."

After Liz and Bertie put in a new section of pipe, they walked back to see if they had taken care of the problem. Water was splashing into the rock bowl for the first time in many years.

I was full of questions, but the orchestra was tuning up and we took our seats. From the wind section, Bertie looked over at me and shrugged helplessly, as if to say, "I couldn't explain it anyway."

The following week, Liz and I hiked up to check the intake screen again. The narrow trail zigzagged back and forth through thick brush, bare dirt studded with cactus, and patches of loose shale. The cave was set like a giant geode in the rimrock. Water dripped slowly from the jagged ceiling, pooling in a deep hollow before flowing through the screen and into the pipe. Liz reached into the water and pulled a few twigs and leaves away from the screen. She pointed out the route the pipe took from the cave, down the steep hillside, and across the pasture to the old homestead site at the base of the rocks.

I asked Liz if she had ever dowsed for anything before.

"No," she shrugged, "but I figured I could do it. Comes from my mom's side of the family. I have an uncle, lives way up in the mountains. He casts out demons for some little church up there."

I remembered meeting this old uncle at the Burns family reunion. He was wearing a floppy leather hat and he tipped his head back to look me in the face with a piercing gaze.

Dowsers describe the process of finding running water underground as a kind of picturing things in their mind—another kind of seeing. I have always prided myself on my powers of observation, but at the ranch I often feel blind compared to Liz. She can pause in mid-stride and pluck a small arrowhead, nearly invisible in the dirt, out of an embankment. She can see cows on the bluffs where I see only rocks. One day in early May, Liz told me about the Nez Perce women who used to come in the spring to dig for couse and camas roots on these hills. Liz used a stick to dig up a root from a plant like a small lily. She tried to explain how to distinguish it from the similar but poisonous death camas, but I couldn't see the minute differences. I would've gone hungry rather than rely on my ability to tell them apart.

I had heard that dowsers always choose a "potent object" to use as a pendulum, and I wondered why Liz had used Gina's wedding ring.

"Your mom must have loved Dooley a lot," I ventured, imagining a love so powerful it could find water underground.

Liz snorted. "My dad always said it wasn't him she fell in love with—it was this place."

Below us, the heat shimmered over the meander bar. It was all so hot and bright—the grass, the silt-glazed stones along the river's edge, the metal roof of the barn—but the river was a cool Popsicle blue. A small boat cruised upriver, the buzz of its motor trailing at a distance. From here the cabin was so small I could cover it with my thumb. The tiny flag fluttered in the breeze. I wondered how Gina could have left the ranch so often, abandoning her flower garden, her child, and this place she loved.

The grinding of the truck sounded in the distance, and I climbed back up to the cabin to get ready to leave. I lifted the spotting scope to take one last look at the kestrel, now watching for mice from the cupola on the barn. I folded the cot and put it back in the broom closet. I dumped the water out of the cooler, sluicing it across the dusty surface of the porch. I closed the door and shoved the padlock through the rusty hasp. I walked around Ram's Head Rock, looking at petroglyphs. It was so quiet I thought I could hear grains of sand sloughing from the rock. In the shallows, a heron stood frozen in profile, as still as the figures pecked in rock. At the sound of the pickup coming down the hill, it lifted slowly and flew out over the water. I stood watching it until the truck swung around the corner of the barn with a rattling bang, dogs barking, horn honking.

I put a finger under the collar of my T-shirt and pulled out the key. I always left Ram's Head reluctantly. I wanted to hang onto something from the place: china from the old shack, a smooth rock that fit my palm, an arrowhead. I wanted to take a picture that captured the evening light lingering on Craig Mountain. I wanted to know the name of every flower in the pasture. I felt like a child with an arm-long Christmas list. I want, I want . . .

My fingers closed over the key; better to drop it into a swirling eddy and let it be taken down to the river bottom than to relinquish it now. I longed to feel it always around my neck, a small reassuring weight, warm against my skin.

A few weeks after that night in the cabin, I dreamed the same dream two nights in a row. I woke to the groaning, screeching sound of wood against wood. Hundreds of logs were piling up at the bend in the river at Ram's Head, just like the old log drives on the Clearwater River that ended decades ago. Old-growth trees were being logged in Hells Canyon and sent downriver under cover of darkness. Behind the log jam, the water rose higher and higher. The first time I had the dream, I dithered about in confusion: Should I stay or go? If I go, what should I save and what should I leave behind? The second night I felt calm as I shone my flashlight on the water rising against the rocks. This time I knew what to take with me: the spotting scope, the sleeping bag, the key.

TWELVE

Everyman's Right

A few miles outside Sitka, Alaska, a narrow road leads down to a rocky beach and a cabin perched just a few yards above the high-tide line. When my grandparents left their island home, they moved to this secluded cove, at that time connected to the highway only by a steep trail. Scotty took to greeting strangers with his shotgun, demanding to know who they were and what the hell they wanted on his beach. The trail became known as "Shotgun Alley," a name it retains today, although the path is now a paved street. My grandfather would have been right at home in Idaho, where landowners are dead serious about their private property rights.

A take-off on the old Woody Guthrie song "This Land Is Your Land" is popular around here. We first heard the song from Bertie, who has a boundless supply of jokes and stories picked up on her odd jobs in the canyon. We were camped on Craig Mountain on a moonless night in late summer when it was too dry to light a campfire. The boys begged Bertie for something to entertain them as they sat on their sleeping bags in the pitch dark. She soon had them singing along to:

This land is my land, this land ain't your land.
I've got a shotgun, and you ain't got one.
You better get off, or I'll blow your head off.
This land belongs to only me.

The song could well serve as the Idaho state anthem.

When we decided to leave California and were considering my sister's invitation to come to the inland Northwest, we got out the map and studied the area. We saw few towns and large swaths of national forests marked in green. If we had been able to zero in on a map of the county where we eventually bought a home, however, we would have seen precious little green. Whitman County has less public land than any other county in

Washington State: over 95 percent of this sizable county is private property (most of the rest belongs to the local land grant university).

From the top of Kamiak Butte, twenty miles or so northwest of Pullman, the entire Palouse is visible. Hills like crested waves roll from the breaks of the Snake River to the mountains behind Moscow. To the south a blue smudge marks the rim of the Lewis-Clark Valley. To the north is Steptoe Butte, near the foot of which Percival Burns had his hog farm in the 1870s. With the exception of these two buttes, everything within view—hundreds of square miles—is private land.

Growing up in the South Bay, I took the beach for granted. It seemed natural to have a large expanse of public land within walking distance of home. A private beach seemed an oxymoron, yet there were cities at the north end of the bay that allowed landowners to build fences right down to the water. Redondo Beach would have had restricted beach access, too, had it not been for the early city administrators who had the foresight to buy up the private beaches.

It was a good thing Southern Californians had the beach to head to for recreation, because the L.A. area has always had fewer parks than most other urban areas. The few parks we had in Redondo Beach were so little geared toward children that they did not even figure into our childhood landscape. We ventured into the formal, hushed confines of Veterans Park—enclosed by tall hedges that damped the sound of waves just yards away—only to check books out of the tiny Carnegie Library. When we wanted to play, we went to the beach, the school playground, or an empty lot.

Fortunately, the patchwork pattern of growth in Los Angeles had left a large number of empty lots. We could launch a packing crate raft on the edges of a seasonal pond or watch our brothers do wheelies on bare sand hills. The words "Private property!" were just verbal darts we threw at kids who intruded on a lot we had claimed. They responded with the obligatory "Free country!" We routinely cut across other people's yards, hopped fences, and picked lemons from neighbor's trees. For the most part, adults tolerated this behavior. As long as there was no vandalism involved, trespassing was considered a minor offense. But in Idaho, not even kids get away with trespassing.

I might have understood this angry defense of property if it were a cherished need for privacy. Yet the same rule applies to land miles from anyone's residence: croplands, creeksides, woods. "ASK FIRST! before you hunt or fish on private land!" is the hands down favorite bumper sticker in Idaho. Yet you can drive for miles without seeing a farmhouse, and there is seldom anyone in the fields. Whenever I've gone up to a house to "ASK FIRST!" (I imagine you have to shout these words), I've braved a sea of barking dogs and, inevitably, found no one at home. The only person ever to answer my knock was an elderly woman who obviously was afraid to open the door to a stranger, and who appeared relieved that I only wanted to ask if I could walk through the draw behind her house up to the highway.

I could understand complaints about people cutting fence wires and vandalizing property, but I had trouble believing that landowners would object to the mere presence of a person on their land. It took a couple of unpleasant encounters to convince me. The first time was when Ambrose and I were following animal tracks in the snow and wandered too close to what we thought was an abandoned greenhouse; the second, and last time, was when I was so hungry for the sound of running water that I climbed a fence to stand beside a creek, bringing a man charging across a quarter of a mile of wet grass to yell in my face.

I told myself that if I ever own land someday, I'll have a more benevolent view toward people from town craving a bit of what I have—solitude, space, silence, water, trees, a good rock to sit on. I wondered if I weren't just kidding myself; maybe one's outlook automatically changed with the acquisition of land, and I, too, would become a mad landowner, crying "It's mine, all mine!"

My great-grandparents in Sweden enjoyed the kind of freedom to roam that I longed for in the West. *Allesmanrätten*, "everyman's right," allows citizens access to the countryside. As long as you don't vandalize the property, litter, or generally make a nuisance of yourself, you can fish in someone's pond or hike across their property, pick berries in a private wood, even camp for a night.

The system's success in Sweden may depend on the unified culture many feel is threatened by the European Union and the new bridge to Denmark; the Swedes envision their private land being inundated with foreign immigrants and day-trippers oblivious to the unwritten rules that have allowed *allesmanrätten* to exist for centuries. For the same reason, here in the West, where there are people of many different backgrounds and wildly divergent opinions, something like *allesmanrätten* probably wouldn't work.

Without access to private lands, most westerners' experience of nature is limited to state and national parks, national forests, and Bureau of Land Management lands. A huge proportion of the West belongs to the state or federal government, so there is no lack of land or (despite recent efforts to close some roads) access. But there is a qualitatively different type of experience to be had on private versus public land.

Public land comes with a vast number of rules, all spelled out, posted, regulated, and enforced through the assessment of fines and, increasingly in recent years, user fees. No littering, no picking flowers, no cutting firewood. Stay on the trail, fires in fire rings only, dogs must be leashed. Trail permit required. I grew up abiding by these rules, and I taught my children to follow them as well. It was part of being a good citizen and a responsible camper. On private land you are, for the most part, free from such rules.

When I met the Burns family, I was both appalled and delighted at how they broke all the rules. Their litter was not a gum wrapper or two: it was whole cars and refrigerators. They chopped down trees for fence poles. They let their dogs run loose. They burned their garbage, something L.A. residents had not been allowed to do for decades. They could run machinery at all hours, yell as loud as they wanted, let their dogs bark and not worry about the neighbors or noise ordinances. I understood for the first time the fierce protection of private property rights in the West. Who would want to give up that feeling of being able to do whatever the hell you wanted to do on your own land?

I had had a brief taste of that freedom on childhood visits to a farm in West Virginia. During the four years my father's company worked on a Department of Defense contract, we lived in a sedate Washington DC suburb. In the white-gloved formality of the Kennedy era, my parents were known as "nutty Californians" because my father drove a foreign convertible and they barbecued in the rain on muggy summer evenings.

Occasionally we would escape to a small farm owned by two "bachelor friends" from my dad's work. They had dug up a coffin on their property and kept it in their living room. It was about the length of their couch and made a convenient coffee table. They told us there was a skeleton inside. (Many years later my father assured me this was true.) One of the men came from a long line of New England eccentrics, and every year he had clams and seaweed flown in from Maine for a clambake. They dug a trench, lined it with seaweed, and filled it with clams, corn, and potatoes. Then they built a fire over it and raked the coals around with metal garden rakes. I was amazed they were allowed to do these things: dig holes in the yard, put a dead person in their living room, play with fire. Apparently, if you got far enough away from everyone—say, at the end of a dirt road in the mountains of West Virginia—you could do anything.

Rural landowners and public land users in the West seldom see eye to eye on land issues. Many of the people using state and federal lands for recreation are middle-class baby boomers from the suburbs. In the 'burbs, everyone is expected to play by the same rules. Suburbanites are the world's champions of fairness. If they can't light fires and cut down trees, they don't want you to either; who owns the land is irrelevant. Ranchers and farmers, accustomed to the vagaries of weather and the market, don't expect life to be fair. Things come down a certain way, and there you are, you deal with it. Sometimes you get the short end of the stick; that doesn't mean you go around breaking everybody else's stick off the same as yours.

One day Liz and I were sitting high up on the cabin porch at Ram's Head when a brown arm appeared around the side of a rock, at about our eye level and less than eight feet away. We watched as the rest of the figure slowly appeared, crabwise, around the corner. It was a sweaty young man wearing only a climbing harness that drew his shorts up around him like a diaper. He was so startled to see us he almost lost his grip.

"Oh," he said, like everyone else we met at Ram's Head, "I didn't know anybody lived here."

Liz informed him that the rock formation, although it appears solid, is actually quite crumbly and unsafe. He apologized profusely and hitched down and out of sight around a boulder. I thought maybe Liz had just been feeding him a line.

"Is it really that dangerous?" I asked.

She laughed, "Yup. I oughta know. I used to climb all over this place as a kid."

What was remarkable about this encounter was that the words "private property" were never used. Liz doesn't always tell people to leave when she finds them trespassing. As long as they're not doing something dangerous or damaging, she often lets them stay. But she makes it clear that someone is watching over things here and they better not get out of line. She carries a gun with her at all times. Although she's never had to use it, she leaves it in clear view on the dashboard whenever she stops to question a hunter on her property.

People are liable to wander onto the ranch from any direction. The west side is accessible by water, and much of the rest of it is bordered by state land. The state gradually acquired the majority of land on Craig Mountain as ranchers died or called it quits and moved to town. Less than 5 percent of the mountain is left in private hands. The Idaho Department of Fish and Game manages the land north and south of the ranch; to the east, it's the Idaho Department of Lands. Other parts of the mountain are administered by the Bureau of Land Management and the U.S. Forest

Service. The Nature Conservancy owns the old Madden Ranch (now the Garden Creek Preserve), and the Heistemann Ranch was bought by the Nez Perce tribe.

The Burnses' access to their summer pasture is dependent on maintaining an amicable relationship with the Department of Fish and Game. The summer pasture is high on Craig Mountain on a part of the ranch separated from the riverfront property by a section of Fish and Game land. After climbing a series of steep switchbacks, the rock-strewn track straightens out for a short distance before climbing over the top of the ridge to the scattering of houses near Lake Waha.

From the road, you can look down at the land compressed into ridges and draws on either side of the creek, and tucked between the folds is a glimpse of the river, five miles off. Instead of the roar of passing jet boats on the river, there is only the sound of wind in the trees. Even the dust kicked up by the truck smells different—of crushed pine needles instead of sagebrush. *Waha*—a mere breath of a word, like an exhalation of delight—is the Nez Perce word for "beautiful."

The Burnses take the road that runs up to Waha whenever they need to get a load of hay or take their cows to auction. It takes over three hours to get to Lewiston, less than twenty miles away as the crow flies. After a strong rainstorm, the creek can wash out whole sections of the road. Inevitably, the state agency takes its sweet time fixing it, according to some kind of bureaucratic timetable that the Burnses find infuriating. Relations are further strained by the fact that both cows and hunters have a tendency to wander: the game warden is quick to complain whenever any cattle stray onto public land, and Liz is just as unhappy when hunters trespass on the ranch.

Much of Liz's time is spent fixing fences or chasing cows off state land. It's a tough country to keep fenced. The fences march up steep ridges, across talus slopes, and through mountain meadows that turn into bogs every spring. Fences get knocked down by rock slides and broken down by determined bulls. Elk get their antlers tangled in the barbed wire and may rip out a whole section trying to get loose.

The Burnses' fences are maintained better than many of the fences that divide property on Craig Mountain. After most of the family ranches were sold to out-of-state landowners, large areas of the mountain were left unattended. For years, hunters, snowmobilers, and four-wheelers felt free to come and go on the back roads. But recently they have encountered more and more roads closed by a locked gate or an earthen berm. The Garden Creek Preserve is open to the public, but you have to get there

Hay barn, Waha. Photo by the author.

on foot, by horseback, or by boat. The Nature Conservancy has closed the road into the preserve to reduce the spread of weeds caused by seeds caught in vehicle tires and wheel wells.

The Department of Fish and Game has started to use road closures to cut down on poaching. Without the means to get in and out of the woods quickly, most poachers don't bother. The department also has declared certain areas off limits in winter because they've had problems with snowmobilers chasing anything that moves, especially deer and elk. Now that there are few timber outfits on the mountain, many of the old logging roads are no longer open or maintained. The various agencies on Craig Mountain are trying to improve the network of trails to encourage hiking, cross-country skiing, and bird-watching rather than activities that require intensive road use, but the country is so rugged—and in the summer, so hot—that few people want to get out and hike, let alone pack out a bagged elk on their back.

I had heard Liz's opinion of Craig Mountain: it is logged over, hunted out, and entangled in a snarl of governmental regulations. But when I first started coming to the ranch, the mountain seemed wild and mysterious. I couldn't even put a finger on what Craig Mountain was, where it started and ended. After a steep climb to the eastern rim, the mountain fizzles out. The jagged peaks of the Selkirks or the classical volcanic cones of the Cascades are as simple and comprehensible as a child's drawing. But Craig Mountain is hidden by the walls of the Snake River Canyon until one last turn shortly before Burns Ferry, when suddenly the canyon opens out just enough for a forested shoulder to appear. Seen from this perspective—low, close up, like a child's distorted view of an adult—the mountain appears all bulges and strange folds. The shapes change with the angle of the sun; at different times of year, the entire mass seems to shift, pressing forward at some times, receding at others. In winter, Craig Mountain looms over the Burns Ranch, dark and massive, sending flash floods of cold air down the side canyons as soon as the sun sets. Seen through the heat-shimmered air of summer, the forested rim looks very high above the river, cool and out of reach.

Even after I had made several trips up the mountain by foot, horseback, and truck, I was still puzzled. Finally, by poring over maps and catching glimpses of it from the air, Craig Mountain started to come together in my mind as something resembling a mountain. The mountaintop plateau extends north-south for over twenty-five miles, from the tiny community of Waha to the ghost town of Zaza. The mountain is sharply defined on three sides: to the west by the Snake, to the south by the Salmon River, to the north by the Lewis-Clark Valley. (In a state with an average elevation of 5,000 feet, Craig Mountain is little more than an average mountain; what makes it seem higher is its proximity to the Lewis-Clark Valley, which, at 750 feet above sea level, is the lowest point in Idaho.) On the fourth side, the mountain's edge is harder to find. The uplift slopes down gradually to the east, eventually merging with the Camas Prairie at around 3,000 feet, near the center of the Nez Perce Reservation.

Craig Mountain was once part of the Nez Perce tribe's vast traditional lands, which covered sizable portions of what is now Idaho, Washington, and Oregon. Through a series of treaties, the Nez Perce reservation shrank until it is now the shape of a small casket, with one corner resting against Waha Lake, just above the Burns Ranch. It is well known that Chief Joseph lost his stunningly beautiful country around Lake Wallowa (unnecessarily so, since it is high and remote and could have been set aside with just a minimum of forethought), but chiefs Toolhoozote, White Bird, and others also lost their country along the Snake and Salmon River breaks and the Camas Prairie. The boundaries of the reservation were drawn to preserve the land of the cooperative treaty Indians based along the Clearwater River.

As if the Nez Perce had not sacrificed enough, in 1893 the federal government decided that the assimilation of Indians into white society would be easier if they were individual landowners. With 160 acres allotted to each head of family and single person over twenty-one, it was calculated that the Nez Perce needed only 263,000 acres. "Surplus lands"—over two-thirds of the reservation—were to be opened to settlement. It was a bonanza of over 500,000 acres for settlers who flocked to the area in anticipation of the opening. Like the "Sooners" of Oklahoma a few years before, the prospective settlers camped out in the counties surrounding the reservation, making forays onto the Camas Prairie and along Craig Mountain to look for promising sites. When the cannon was fired in front of the land office in Lewiston at noon on November 20, 1895, the local newspaper proclaimed it "the death knell of the great Nez Perce Reservation."

Over the years, many Nez Perce landowners sold or leased their lots to whites. Now only a small portion of the reservation is still owned by the Nez Perce—13 percent, according to the tribe's own figures. And those sections have been divided and subdivided until they often have as many as twenty owners, making improved management, sale, or development of them a near impossibility.

Under an agreement with the state, the Nez Perce are allowed to hunt and fish on state and federal lands on the mountain without being subject to Idaho hunting regulations. The tribe also has its own land on the mountain, including a parcel bought with mitigation money received from Bonneville Power in compensation for flooding their hunting territory when Dworshak Dam was built on the Clearwater River. The tribe does have its own proscriptions about hunting in particular areas and at certain times of year, but some tribal elders complain that many of the younger members of the tribe don't follow the old traditions about how and when to hunt. Even Liz, who generally respects Nez Perce traditions, says, "When a moose wanders onto the mountain, you can be sure it won't be there long; the Indians will get him."

In my effort to understand Craig Mountain, I decided to talk to someone at the Idaho Department of Fish and Game (IDFG), even though Liz warned me that "you can't trust a word they say." Dooley has no patience with state employees either. "They don't listen to you. Just ignore you and go about doing things their way." I discovered there is frustration on both sides. The Fish and Game wildlife biologist said the Burns family doesn't understand that "you have to go through the usual channels to get anything done. Dooley doesn't even try to deal with us anymore. If he has a complaint, he just fires off a letter to his senator."

Ranchers are under the gun everywhere in the West, but they come under particular scrutiny on Craig Mountain because it has a significant number of rare and endangered animal species and is a "hot spot" for rare plants in the state. The mountain is able to support a rich assortment of wildlife and plants because it encompasses everything from arid grasslands along the river to upland forests.

The impact of grazing is a touchy subject for the Burnses, because cattle have been blamed for the changes in the grassland environment in the canyon, as they have been throughout the West. Intensively grazed land (which, in the canyon, means anywhere flat) has been invaded by non-native plants such as yellow star thistle, leafy spurge, and the ubiquitous cheatgrass.

The Nature Conservancy has removed most of the cattle from their land, primarily because they transport weed seeds, spread them around, and trample them into the soil. The IDFG intends to follow suit. To support their theory that the area was not meant to support grazing animals, IDFG scientists point to the native grasses, which grow in bunches separated by bare rock or an easily disturbed crust common to desert soils. If animals were meant to graze on Idaho rangeland, they claim, then the grass would be turf grasses that form the thick, almost indestructible sod once found in the Great Plains where buffalo roamed.

When I ran this analysis past Dooley, I was not surprised to hear him disagree.

"When Lewis and Clark came through in 1805," Dooley said, "there weren't many big mammals in the mountains. They practically starved. We had sheep summering in the mountains for over fifty years. After sheep made the turnover, the mountains began to produce high-quality feed, and then you got the deer and elk. But if you take the big animals off, the grass cycle dies—no more manure. The pasture turns to shrubs. Now sheep are out and the big mammal population is dropping."

There is no question that elk populations throughout north central Idaho have dropped drastically, and most hunters are strong advocates of any management practice that will benefit the elk, such as decreasing the number of predators and logging to create the open areas that elk favor.

Sheep are generally considered to be hard on rangelands because they nibble grass almost down to its roots, a problem made worse by the fact that they are often raised on lands that are marginal in the first place. Dooley takes the contrary view that sheep are healthier for the land.

"Cows are lazier," he said. "They stand in one place and chew. Sheep are ambitious. They think the grass is always greener. They have more energy, so they travel more."

It's a stretch to try to imagine sheep as ambitious creatures: upwardly mobile herd animals; yuppies with hooves.

At one time, I would have agreed with environmentalists who want all grazing animals off public lands in the West because of the damage they cause, especially to areas along streams. Thornbush Creek runs through the Burnses' summer pasture, yet it does not look anything like the pictures I have seen of trout streams ruined by cattle grazing. It is the only creek for miles around that produces significant salmon and steelhead runs (the others are too warm or shallow, run only seasonally, or are too degraded by mining and ranching activities).

The ghastly photos that often appear in environmental publications show eroded banks trampled bare of any vegetation. Before-and-after pictures show how the streambeds have become progressively wider and shallower (and thus slower and warmer) until they are unable to support trout, which need cold, fast-flowing water. But Thornbush Creek is well shaded by trees, the banks are stable, and the water is clear and fast moving. The cows don't spend a lot of time along the creek, partly because Liz keeps them moving from one fenced section to another, and also because there are plenty of other small springs where the cattle can find shade and water.

Ranchers such as the Burnses have shown that grazing doesn't have to mean devastation to the environment. I used to think that stock animals have no place in our forests and on our rangelands. But I've come to value the link between work and the land that grazing animals represent. I understand now my father's deep respect for people who got their hands dirty making or fixing things. He was a white-collar worker, an engineer, but he was influenced by early memories of watching his grandfather laying brick. On a visit to the Palouse, he took an outsized delight in sitting in the fanciest restaurant in Pullman and having his view blocked by a hay truck hauling its load through town.

"That's what I miss about old Redondo," he said, "the sense of it being a real working town." As a resort town, Redondo Beach has always been concerned about maintaining a certain image. Back when Redondo was the only deep-water port serving Los Angeles, as many as two dozen ships a day took on lumber and other freight brought by train from L.A. But once tourism started to catch on, city boosters complained that seaside commerce wasn't compatible with the image of genteel leisure they were trying to convey. In other words, no one wanted to see anyone actually working.

Until World War II the town retained some semblance of being a working town. Soon after the war it started its decline into a bedroom community. The area lost its agricultural and fishing base and much of its oil production. Then industry went: the pottery works, factories to make glass and boxes, and plants to process salt and silk. Next to go were the small businesses that anchored the downtown: meat markets, hardware stores, shoe repair shops. Finally, unsightly utilities such as the sewage treatment plant were moved to less visible locations. By the 1960s there were so few signs of the actual workings of the town that we might as well have been living on Disney's Main Street. In recent years, this air of unreality has been heightened by a rebirth of the resort industry centered around the pier.

The Snake River Canyon and some parts of Craig Mountain are real working landscapes. In this day and age, working landscapes are becoming almost more scarce than true wilderness. Part of the Snake River Canyon's appeal to visitors is the long intertwined history of people and nature that is apparent along the river, from archaeological and petroglyph sites to copper mines and limestone quarries. Tourists are intrigued by seeing how Nez Perce bands, Chinese gold miners, and white homesteaders overcame the challenges of living in this rugged and remote location.

Dooley said the passengers he used to take on boat tours of Hells Canyon loved nothing more than exploring abandoned homesteads and mines and digging through old dumps. When the Hells Canyon National Recreation Area was formed, a misguided attempt to restore the canyon to a more natural state included eliminating any evidence of recent human occupation. The Forest Service burned down most of the buildings and ferried all the trash and scrap from the old dumps by helicopter to a central pit, where it was buried well out of reach of any sightseer who might be tempted to take home a biscuit tin as a souvenir. Only in recent years has the Forest Service come to recognize the historic value of the ranches and mines and started taking steps to preserve what little is left.

The Fish and Game fisheries biologist told me that Thornbush Creek is "pretty good habitat. It's not pristine, it's been worked, and in some reaches, the presence of cows is a problem. But the Burnses manage their cows better than the owners of some other creeks I look at on Craig Mountain."

Dooley regards the presence of cattle as a plus for the creek. "There's no doubt but the livestock being in the creek increases the fish food. The manure and whatnot makes more worms and bugs."

There can be too much of any good thing, even a cow's "whatnot," but it's true that many western streams are starved for nutrients now that they are no longer lined with the decomposing bodies of spawned-out salmon. The creek produces a healthy number of rainbow trout, at least in part because of the Burnses' conscientious management. But there are some who feel they could do an even better job.

"Some of these wildlife preservation deals and fish people wanted to take over Thornbush Creek," Dooley told me. "They wanted to put in a fish-counting thing up there and they wanted to do some channel work. If you'd a got 'em in there, you wouldn't dare turn any water out into the field to irrigate. They'd have been right there after you. They'd be up there: 'Hey, you can't let a cow in the creek here, you got to fence that.' When you fence a crick off, why, it grows up full of blackberries and stinging nettles. You can't get to it. You got to crawl on your belly up the crick. I just fought it until they finally gave it up. I'm not much in love with government regulations."

Government regulation, along with the cooperation of ranchers and other landowners, might have addressed one of the worst threats to ranching in recent years: the invasion of alien plant species. Coming from California, where almost everything and everyone is non-native, it took me a while to recognize the seriousness of this issue. Most of the types of vegetation commonly associated with Los Angeles are actually imports: tropical palms, iceplant, jacarandas, birds of paradise, even the ubiquitous eucalyptus trees. The native chaparral is such a fire hazard that no one wants it around residences. Southern Californians who feel threatened by the deluge of illegal aliens would find the idea of an invasion of alien plants laughable.

In rangelands throughout the West, ranchers and rangeland managers have watched the relentless spread of these weeds with growing horror. Nevertheless, they have not reacted with a coordinated response. Public land managers found that weed suppression was a low priority with the public. Most people neither knew nor cared about weeds, and won't, more than one range scientist has noted, until popular hiking areas became impassable, thorny wastelands. Weed eradication programs were underfunded and under attack by environmentalists. Ranchers saw government agencies taking little action and wondered why they should tackle the weeds if they were just going to spread onto their property again from the public lands.

Plagued with other problems (low beef prices, growing regulations, the debate over grazing allotments), ranchers tried to ignore the problem with weeds, even as it began to cut into their profits. A local agriculture extension study on yellow star thistle noted that area farmers and ranchers often ignored a dip in profits that would have set off alarms and prompted immediate action in any other business. (Almost as revealing as the study itself was the fact that in the six years the study ran in the 1980s, almost 40 percent of the farmers and ranchers died or left the business.)

Pursuing the topic further, I checked out the county's agricultural extension website. Clicking on yellow star thistle, I watched the blank outline of Nez Perce County fill here and there with dots of yellow. The dots multiplied, merged, and then suddenly mushroomed across a large area of the map. This, in slow motion, is what the Burnses have watched take place on their own land in recent years. Liz estimates they have lost half their pastureland to invasive weeds, primarily yellow star thistle.

Of all the non-native species in the West, cheatgrass is the most widespread. Cattle will eat cheatgrass when it is young and green, but when it matures, it develops barbed seeds that can be dangerous to animals, working their way into a dog's ear or the back of a cow's throat. After a walk across the pasture at Ram's Head, my socks and shoelaces are usually covered solid with cheatgrass seeds and burrs. (Liz's solution is simple: don't wear socks.) But chukar, an introduced species of game bird, depend on cheatgrass for food, and some scientists even claim that without the green-up of cheatgrass after the fall rain, deer wouldn't go into winter with enough nourishment to last until spring.

There is nothing redeeming about yellow star thistle. The spiked flower heads are so sharp they leave deep scratches and welts on my legs. (I don't have to ask what Liz's solution is: don't wear shorts. Not an easy concession to make in one-hundred-degree weather.) The spines can damage a cow's eyes as it lowers its head to graze, and a toxic chemical in the plant can kill a horse. Nevertheless, cattle eat the young thistle in the spring, and ranchers say that if you take the cattle off the land, the star thistle population will explode. Environmentalists say that if it weren't for cattle, there wouldn't be a problem with star thistle in the first place. Cows disturb the soil, providing an opportunity for the invasive weeds to become established and crowd out the native grasses. They go round and round about it, and nobody knows whom to blame or what to do.

Ranchers blame environmentalists, particularly the Northwest Coalition for Alternatives to Pesticides (NCAP), who were so adamantly opposed to any use of chemicals that they initiated a lawsuit that resulted in an injunction against all herbicide use on Forest Service land in Oregon and Washington. In the five years it took for the Forest Service to revamp its vegetation management plan and settle the lawsuit, yellow star thistle took over thousands of acres of grassland. Now any spraying would have to be done on a much more massive scale.

Soon after we moved to the Palouse, I became one of the founding members of an environmental group whose main purpose was to fight the heavy use of pesticides and herbicides by area farmers and the local school and park districts. Our group supported NCAP unequivocally. I first became involved in the issue after I wrote a letter to the editor protesting the farmers' use of a particularly potent organophosphate, called Di-Syston, to fight the Russian wheat aphid that had recently appeared on the Palouse.

The environmental group succeeded in getting the parks to post signs after they sprayed, but we failed to get the school system to rely less on pesticides and institute an integrated pest management program. In the end, I don't know what impact we had on the farmers. At the time, I was sure we were working for a good cause. I felt justified in being a "chemophobic environmentalist." I worried about my children playing outside when there were crop dusters spraying the fields just a few hundred yards away, and I hated dragging them away from the park when I smelled the still-wet Roundup on the grass.

I used to flinch whenever Dooley got out his backpack sprayer and went out to squirt weeds. And at one time I wouldn't have been caught dead pricing 2,4-D at hardware stores in Lewiston, as I later did with Liz. But now, looking at the endless acres of yellow star thistle blooming on the ridge across from Ram's Head forces me to reconsider the issue. Every time as an environmentalist I just say no to glyphosate, picloram, dicamba, and 2,4-D, I'm allowing another thistle to let loose thousands of seeds on the side of Craig Mountain or another spotted knapweed to send its long taproot down into my beloved Snake River Canyon.

The Department of Fish and Game is taking a wait-and-see approach to yellow star thistle. They're watching the Nature Conservancy's Garden Creek Preserve to see which of their experiments in weed control will pay off. The conservancy is trying everything in the book, from hand-pulling to pesticides, burning off affected areas and covering them with black plastic. The plastic worked pretty well, but the elk kept slipping on it; one was even seen sliding on its belly down a steep hillside.

The grassland is not the only ecosystem in trouble on the mountain. When Percival and Odalie Burns first settled on Thornbush Creek, the top of Craig Mountain was a parklike setting of open grasslands with widely spaced ponderosa pines. Over the last hundred years, the grass has been overtaken by brush, and the ponderosas have been largely replaced by thick "doghair" stands of fir and lodgepole pine. The ponderosa pine savanna has become the most endangered type of forest, not just on Craig Mountain, but all over the West.

Who is to blame for the change in the forest and what to do about it has been a matter of debate between the various landowners on the mountain. After decades of logging and suppression of wildfires, the grassy understory has filled in with brush, providing feed for deer, who eat leaves, but little for elk. Loggers early in the century practiced high-topping (taking the most valuable timber only), cutting many of the old ponderosas. More recent logging operations have clear-cut wide swaths of forest.

Without the ponderosas, the mountain has reseeded itself with firs. Under natural conditions, fire would have killed off many of the fir seedlings, but sheep and cattle have eaten the grass that would have supported a ground-hugging fire, and the catastrophic fires likely to burn now are quickly suppressed. The tree habitat has gradually changed from sun-loving ponderosa pines to shade-tolerant firs, which are more susceptible to disease, parasites, and fire.

The Idaho Department of Fish and Game wants to return the forest on Craig Mountain to its pre-settlement state. Even I find this hard to understand. It's not a wilderness and shouldn't be managed as one. The mountain has been heavily altered by humans, from the Indians who probably set fires to clear the understory, to settlers who logged and raised sheep, to the well-intentioned game department that introduced chukar.

Even the department admits that their intensive management plan might be hard for many people to swallow.

A 1996 report states, "In the short term, relatively intensive disturbance to the forests will have to be understood and endured in order to accomplish long-term goals and benefits to wildlife, as well as to the general public." The Burnses may feel vindicated to some extent by the IDFG's recommendation that "light livestock grazing" be included in the reforestation plan, to reduce both fire hazard and competition from other vegetation. But the department's scorched-earth policy that combines unsuppressed burning, "sanitation logging," herbicide spraying, tilling, manual scalping, and scarification—in some areas even removing the sod—sounds like the equivalent of using chemotherapy to halt the spread of cancer. I wonder if they know what they're doing, and if, like chemo, they might just weaken the whole system, or kill the patient by trying to save it.

Dooley has selectively logged his land for years. He has a deep understanding of the life cycle of the trees and the proper way to log to leave a healthy forest.

"There's a lot of false information about the damage done by harvesting trees," he said. "Trees have a life like a person—about seventy or eighty years. When white fir gets about twenty-four inches, it gets a rotten center. Red fir, when they get that big, they get wind shaken. They crack, and the crack fills up with pitch. If you've got a half-inch pitch blade in there, it's worthless for boards. And when they get old, sometime they get attacked by tussock moths. It's like a snowstorm, billions of them, like white flies that suck the sap out. So you got to harvest fir at a younger age than pine. Back in the fifties they cut four hundred acres up the crick. They left all the trees that were ten to twelve inches around standing, so it reproduced itself."

The area that was logged forty years ago is a sheer north-facing ridge above Thornbush Creek. The neighbor who owns it has been talking about selling it for years. Sitting on the grass on the hill above Percival's house, Liz said, "First she said she was going to sell it to somebody for a vacation home. Then she was negotiating with the Nature Conservancy. Finally she sold it to a logging company—probably what she was planning to do all along. She just wanted to jack up the price. I met her kids—real nice people. They would've liked to build a home and raise their kids there, but they couldn't afford it."

I remembered that Liz doesn't even cut down a tree at Christmas. She said she doesn't want a hole in the forest for every year she celebrates the holidays. Instead, she decorates the house with pine branches. One cold

December day I stood high on this hillside and watched Liz cut branches from a small pine on the open slope below. She gathered up an armload of boughs and smelled the fresh pine like a child burying her face in a pile of clean laundry.

Looking up at the mountain, Liz said, "Last time it was logged, I was younger than Ambrose. It's taken my whole lifetime to grow up again, and now some wildcat logging outfit from out of state is going to come in and make a mess of it and then just leave."

I had come to the ranch to say good-bye before moving to the Midwest. It would be close to a year before we could return for a visit, and I felt sick at the thought that I would never again see the mountain as it was that day. I fixed it in my mind—the thick stands of fir, the occasional huge ponderosa towering above, the bare fawn-colored ridge top. The air was absolutely still. I remembered Dooley saying that at this time of year it's so quiet that you can "hear butterflies go by making a whoofing sound."

I lay down on my stomach on the grass, feeling the prick of pine needles and small stones against my cheek. I listened for the sound of butterfly wings, but all I heard was the deafening roar of a helicopter plucking the trees one by one and carrying them over the ridge, leaving the mountain bare.

THIRTEEN

Pining for the West

A year after my father died, we made the difficult decision to leave the Palouse and move to Illinois so my husband and I could both attend graduate school. Our first few weeks in Illinois were a nightmare. We arrived in the midst of the worst heat wave in almost sixty years. It was over 100 degrees, but the high humidity made it feel like 120. In four days, more than six hundred people died in Chicago from heat exhaustion. The air conditioner wasn't working and neither was the water heater. We took a lot of cold showers while we waited for a repairman to become available. At night the hysterical pitch of the crickets was so loud I couldn't sleep. During the day, we drove around doing errands in the air-conditioned car, peering out at this new world we had entered.

On baking hot days on the ranch, it had been a relief to spot a bit of cool green—a stand of pines on the mountains or a mossy patch at the edge of a spring. But here the steaming fields of vitriolic green tired the eye with their intensity. Every time I stepped outside I gasped for breath. I felt out of my element: a dry-land person immersed in a dank environment of slack rivers and green-scummed ponds. I longed for the feeling of height, the solidity of rock underfoot, and air so dry it cracks the skin.

One night I took the dog for a walk. I felt oddly removed from everything around me. The humid night air, the sound of trucks downshifting as they got off the interstate for gas, the soft haze around the lights of the truck stop on the highway: these were the same sounds and sights I had experienced as a child on our cross-country travels. This was a place to stop for gas or, in a pinch, maybe spend the night. We had stayed in many such places, sweating on top of our flannel sleeping bags at the back of the trailer, snug as ship's quarters with four children below in a square, three slung above in hammocks. I would lie awake listening to the mosquitoes bang against

Waiting for the mail boat at Somer's Creek, 1943. Photo by Kyle Laughlin. 99-G-005-08. Historical Photograph Collection, University of Idaho Library, Moscow, Idaho.

the window screens, trying to imagine myself lying on top of a cool sheet on my bed at home, the unscreened window open to the ocean breeze.

No, this place was not, could never be, home. We would soon be out West again, I assured myself. Two years, three tops . . .

My husband had grown up in central Illinois, and before we moved here, I had quizzed him on what the place was like. His vague description of flat cornfields didn't tell me a thing I didn't already know. I pressed him for details. "I really don't remember. I didn't pay much attention to nature until I moved out West," he said. The first time we walked in the woods here, he looked at the trees standing in murky water with thick Tarzan vines drooping from the upper branches. Batting away mosquitoes, he said dispiritedly, "I grew up in a fuckin' swamp."

The prairie here was different from the dry, upland prairies along the Washington-Idaho border. The Palouse had been covered with bunch grass so rich that it supported huge bands of wild horses. Early settlers, who used the land for grazing rather than farming, had to hobble their own horses at night to keep them from running off with the wild herds. But this part of the Illinois prairie grew mostly slough grass, a tough, sharp-edged grass that could cut like a knife. The area had been so filled with ponds and low-lying sloughs that the local bands of the Illinois didn't establish villages but only came here to hunt. Early settlers bypassed the region, fearing malaria. The few people who did settle down did so in the oak groves that dotted the prairies; it was less trouble to clear the woods than to wrestle with the tough prairie sod and potholes that filled with water at every rain. But the woods have been cut down for railroad ties and buildings, the sloughs drained and filled, the prairie plowed and cultivated. Even with all the alterations to the landscape, it still feels inhospitable, with tornadoes in the spring and a constant bone-chilling wind in winter.

It was not only the land that felt unwelcoming. Ten years in laid-back Northern California and another eight in the amiable Northwest had left me unprepared for life in this midwestern city. Champaign-Urbana has a combined population of less than one hundred thousand—hardly a big city. Spokane has three times the population, yet the city is cleaner and the people more civil. Midwesterners struck me as unfriendly, harsh-spoken, and inflexible. My husband, on the other hand, felt instantly at home among the people here. His ancestors were tobacco farmers in Indiana from the 1830s until his grandfather came to Illinois to take a factory job in the 1920s. Richard understands midwesterners in a way I probably never will. The type of speech I regarded as rude, he saw as blunt and no-nonsense. When I complained that the people didn't cut each other any slack, he said he admired how "hard-ass" they are.

Richard's aunt Dorothea, I discovered, is as hard-assed as they come. Over eighty years old, she still insisted on cleaning her own rain gutters and killing her own snakes (even if they were just garter snakes) in the backyard of her small house in Decatur. If she thought you were being lazy or indulging in a "pity party," she would tell you so in no uncertain terms. She became a tough but loving grandmother not only to our boys but to Richard and me as well.

We rented a house on the edge of the small town of Mahomet. Behind the house was a cornfield on the other side of a drainage ditch that we euphemistically called a creek. When we had moved to the Palouse, I had expected to see the fields of wheat rippling in the wind. Indeed, old photos of the Palouse show farmers standing in fields full of shoulder-height wheat. But the old, leggy varieties of wheat have long since been replaced by dwarf strains that grow only knee high. In late summer, the hills are covered with short blond stalks with all the romance of a crewcut. I had entertained similar fantasies about living next to a cornfield. All those plump, juicy ears of corn just waiting to be piled high on a white platter at some Norman Rockwell family supper. But the corn was left standing in the field until the stalks were dry and hollow as bamboo, and the ears became gap-toothed as kernels shrunk and fell out. It wasn't sweet corn, I found out, but "trash corn" that was fed to hogs. It was finally harvested in the fall, leaving behind a wasteland of bent and broken stalks. Huge flocks of starlings came to eat the small piles of spilled corn in the furrows; they settled in the trees around our house, dislodging twigs and leaves that fell like brittle rain.

Sometimes I left the interstate and took back roads home through the farmland. It took longer, but at least I could count on getting a friendly wave or two from other drivers on the mostly deserted roads. I enjoyed driving through small towns with names like a gathering of good old boys: Leroy, Dwight, Homer. The style of homes in these towns and on the surrounding farms was oddly familiar: the plain, wood-frame houses that settlers built in the West were modeled on these. The Burnses, once they arrived in the Snake River Canyon, built a house that must have looked much like the one they left in Missouri. In early photos it looks charmingly, eccentrically out of place—a neat farmhouse plucked, Oz-like, from the Midwest and set down in a rugged canyon.

Richard was eager to show us Chicago. He had fond memories of the city from his teenage years when he used to take the bus into the city for violin lessons. We made several trips to Chicago over the next few months. Emlyn loved all of it—the el, the bright displays in the store windows, the giant slices of Chicago-style pizza. We had to drag him out of the museums, still protesting he hadn't seen everything. I liked Chicago, too. Compared to L.A., a city without a center, the downtown was sharply defined and delightfully compact. I enjoyed being able to leave our car behind and ride the el and walk.

Ambrose was disturbed by the whole big-city experience. The first time we passed a street person, he took the money he had been given to spend on souvenirs and ran back to put it in the man's hat. One time we walked out to the end of Navy Pier. It was a beautiful, breezy day and Lake Michigan was bright blue. Ambrose stood at the rail for a long time looking across the water at the lighthouse in the distance and turning in circles to watch the seagulls as they swooped around him. Walking back to the car, we passed a raucous jazz combo at the outdoor cafe—part of the new complex of shops and tourist attractions that had been recently completed at the base of the pier. He sighed, "Why do they have to ruin every good place with tourist stuff?"

Ambrose was able to find enough "good places" around our house to keep him satisfied. Our semirural location allowed him more room to roam than he had had in Pullman. One of his favorite activities was what he called "water walking"—walking along the creek by himself. Sometimes he packed a water bottle, a snack, and a notebook and rode his bike to the woods to draw or write poetry. He was happy to discover that, unlike the children in Pullman, who were often busy with music lessons or soccer practice, these rural kids led totally unstructured lives. He made friends easily and joined the other boys in playing baseball and building tree houses. All summer long, his legs were a mass of welts from mosquito bites, chiggers, and poison ivy.

As soon as the weather was bearable, I joined Ambrose in exploring the nine-hundred-acre county park a short distance from our house. I was surprised that this landscape, which had seemed so alien and unattractive at first, became a source of comfort to me when the city got me down or I was overcome with homesickness. But it was a different type of nature from that we had experienced in the West. There were no grand views, and even deep in the woods we could hear the bell tower chiming "Edelweiss." If we walked quickly in a straight line, we would come to a road in less than twenty minutes.

We had to slow down and follow the meanders of the Sangamon River. We found small things to enjoy: shiny mollusk shells in the mud, a glimpse of a fox bounding over a pile of brush, the musky hollow full of deer lays (semicircles of flattened grass) where the deer bedded down at night. We watched a snake swallowing a bullfrog at the edge of a pond, a blatting kazoo-like sound issuing from the frog with every squeeze of the snake's jaws. Sometimes Emlyn came with us. The boys swung on vines, threw rocks in the river, or jumped up and down on giant slabs of ice that rocked like teeter-totters.

In our years of visits to the ranch, I had learned that you can't really know a place until you've been there at all times of day and night, in all kinds of weather, and in all seasons. We didn't wait for good weather to go to the park, when picnicking families crowded the lawns and boys with fishing poles lined the banks of the lake. Ambrose and I ventured out on rainy or raw winter days. Often we had the whole park to ourselves. We went at dusk to watch for deer, and at night to listen to the frogs in the ponds. We went to the golf course early in the morning when the snow was unmarked except for the track of a single cross-country skier.

The park was also a good place to watch for birds. Illinois lies smack in the middle of one of the major flyways for migrating birds. We were excited to see scarlet tanagers in the woods and terns diving for fish in the small ponds. One day Ambrose burst into the house, calling for me so urgently that I thought one of his friends may have been hurt. "Mom," he shouted, "I just saw a snowy egret in the creek!"

After school, Ambrose volunteered at a nature center in a city park bordered on one side by gang territory. There he fed dead baby mice, called "pinkies," to the screech owl and snakes, and leftovers from staff lunches to the turtle. While Ambrose worked in the center, I walked through the park, which had more different kinds of trees than I had ever seen in one place outside of an arboretum. I was used to the simplicity of evergreen forests, with subtle variations of pine, fir, spruce, and cedar. Here every tree held something strange and wonderful, from the plate-sized leaves of the sycamore to the spiny baubles on the sweet gum tree. I often left the grassy parkland to walk in Busey Woods, one of the few remaining remnants of the oak and hickory wood called Big Grove that once grew north of Urbana in a swath thirty-five miles long and nine miles wide.

The original prairie hasn't fared much better than the old groves. We had left Idaho, where less than 10 percent of the land is developed, and moved to Illinois, where less than 0.1 percent of the land has been left in its natural state. I was curious to see what a midwestern prairie had looked like years ago. We caught glimpses of it on the roadsides where the highway department has been experimenting with planting native grasses and forbs, but I wanted to see a large area to get some sense of the huge expanse of prairie that settlers had encountered 150 years ago.

We drove thirty miles to a park owned by the university, where we could walk through a fifty-acre restored tallgrass prairie. It was a late afternoon in the early fall. The slanting light lit up the acres of goldenrod, and the plumed seedheads of Indian grass waved in the breeze like small prayer flags. I had often heard that the grass had been tall enough to tie over the pommel of a horse's saddle, but I had had trouble envisioning it—I kept imagining a dense growth like a field of corn, impossible to ride a horse through. But I saw that the prairie was a complex mixture of dense short grasses—little bluestem, pioneer dropseed, sideoats grama—and the tall delicate fronds of big bluestem and Indian grass. Standing in the middle of that small patch of prairie, I felt a curious mixture of delight and outrage that I hadn't felt since I stood beneath the largest of the giant white pines left in a logged-over park in Idaho. That day I had wanted to shake those turn-of-the-century loggers by the collar and ask what right they had to deprive future generations of the sight of an entire forest of these soaring, five-hundred-year-old giant pines.

Hungry for news of Idaho or Washington, I often scanned the paper for stories about the West. In Idaho, we read more than we wanted to about current events back East. We also heard quite a bit about the Midwest, since the media seems to regard the heartland as the place to go to find out what "real Americans" are thinking and doing. But in Illinois, the West seemed hardly to exist. Most midwestern newspapers, even the Chicago papers, ran stories about L.A. and Seattle, but almost nothing about the intermountain West.

One of the few events that warranted news coverage was the widespread flooding in the Northwest that spring, caused by an early thaw after a winter of unusually heavy snows. My sister wrote that the three creeks that ran through Pullman (the old town of Three Forks) had risen to the level of the railroad bridge. Small towns along the Palouse River were six feet deep in water. Concerned about the Burns family, I sent an e-mail message asking how they were doing. Liz wrote that the road was filled with rocks the size of cars and the hog farm near Rogersberg had been washed out, sending six hundred hog carcasses floating down the river. My kids were upset to think they had missed the flooding—the most exciting thing to happen in Pullman in years.

I wasn't the only one in Illinois who missed Idaho. I had become friends with a woman who had grown up on a sheep ranch on the rim of the Snake River in southern Idaho. Six feet tall and rail thin, Marva stood out among the sturdy midwesterners. When they teased her about being a hick from out West, she slipped into an Idaho drawl and told amusing, self-deprecating stories about how naive she had been when she left Idaho for the first time at the age of eighteen. After ten years in Illinois, Marva still didn't feel at home.

One afternoon we sat in her kitchen trading ranch stories. She talked about the long days she spent on horseback, tending the sheep. To pass the hours, she would ride backward or standing up, singing every song she knew just as loud as she could. When she was finished with her chores, Marva used to slip off to a private place where the creek that ran past the ranch house flowed over the canyon rim, splashing into a small pool hidden behind a screen of willows. At night she slept by an open window; in the stillness she heard only the sound of the creek and, beneath it, the low drone of the river.

Marva still mourned the loss of the ranch. Like many family ranches, it had been sold to settle squabbles over joint ownership. At the time of her father's death, Marva was in South Korea serving in the army. Before she could get home, the ranch had been sold and the contents of the house auctioned off.

"Nobody even thought to ask me if there was anything I wanted. I would've liked to have my father's hat. He was probably the only rancher in Idaho who wore a pith helmet." She sat staring out the window, lost in thought. A small plane, coming in to land at the airport, buzzed in the distance. "I never got to say good-bye to the place."

I busied myself rinsing my teacup in the sink for a moment until Marva looked around and smiled. I asked her something I had asked myself many times over the last few months, trying to narrow my answer down to something essential.

"What do you miss most about the West?"

She said, simply, "The quiet."

While we were in the Midwest, I wanted to visit some of the places where my ancestors had lived before starting for California over a hundred years ago. In Kentucky we poked around an old cemetery in the village of Crab Orchard. Many of the gravestones had toppled over and broken, and some had been heaped in an overgrown corner of the churchyard like so much rubble. It was a windy, overcast day in January and the boys were anxious to leave. The place was "too sad," they said. I was bothered, too, by the thought of the people buried here (perhaps some of my ancestors) without names or stones to mark them.

In Iola, Wisconsin, I spent an afternoon in the county seat looking up land records, obituaries, and cemetery records. On old plat maps, it was gratifying to see how much land around the town of Iola had belonged to my great-grandmother's family. With the exception of Uncle Howard's small orchard, this was the last chunk of land my family ever owned. The Taylors' unprofitable farm in the thin, rocky soil of northern Maine had been simply abandoned. The Buford family's prosperous dairy farm on the banks of the Ohio River had been lost in the Civil War. The Johnstons, the Hobergs, the Rhodeses all became city folk eventually, living in apartments or houses with small tidy yards. Think of it—almost 130 years since we've had land in my family. Maybe that's where my great hunger for land comes from.

I asked the librarian if they had copies of the town newspaper dating back to 1895. She said the early issues had been lost in a fire, but I could check with the publisher and see what they had in their morgue. I had to ask twice to be sure I understood: The newspaper that my great-grandfather had established was still being published? It seemed incredible to me that a weekly newspaper could survive in this tiny town for over a hundred years. Times were tough for small newspapers, and even Franklin Johnston's newspapers in the South Bay had gone under in the 1980s. The librarian directed me to the newspaper office down the street.

I bought a newspaper and introduced myself to the publisher. He seemed pleased to meet me and interested in hearing about the man who had established the *Iola Herald*. I had with me a photo of Franklin Johnston and his wife Lenora Taylor, and the publisher asked for a copy to hang on the wall. It pleased me to think of Franklin, in his pince-nez and curled mustache, looking down once again upon the bustle of a small-town newspaper office.

The town records indicated that the Taylor family was buried in the old cemetery outside of town. It was a peaceful little place with a backdrop of mountains and stacks of logs—I could see that Iola had not strayed far from its roots as a lumber mill town. We scattered through the graveyard looking for the Taylor name. My husband called to me that he had found a Henry and Henrietta Taylor. Having come all this way to Wisconsin, I thought I might find nothing more than a few impersonal scraps of information, or catch a glimpse of the past (as we had at Franklin Johnston's birthplace in Fond du Lac) obscured now by gas stations and video stores. But I had had no idea that visiting the graves would be such a moving experience.

When I saw the names of my great-great-grandparents, I dropped to my knees in the grass, tears in my eyes. Puzzled and slightly exasperated, Ambrose asked, "Why are you crying about people you didn't even know?" He was so used to seeing me crying over the dead that I had to explain to him that I wasn't sad. I was happy that could I put my hand on that limestone obelisk and feel a connection to ancestors I had known only as skittery signatures on old documents. I also felt a strange sense of relief. Maybe it's not just land I need, but a family plot—somewhere to lay to rest those family members I've been packing around for so long.

On a walk through the cornfield behind our house, Ambrose and I discovered that the creek had been stopped up by a beaver. It looked like the work of a single animal, pushed out of the woods beyond, perhaps, where a new subdivision was going in. At a natural bottleneck in the creek where trash accumulated, the beaver had built a fine dam. Between the sticks packed with mud, we could see the end of a red plastic sled, a child's wading pool, the bomblike shape of a fire extinguisher, a bicycle wheel. I admired the beaver's hard work and ingenuity at the same time that I felt sorry for it.

We, too, had been trying to make the best of the situation here, but sometimes the effort of imagination required to turn a drainage ditch into a creek, or a wooded park into a real forest, got to be too much. The novelty of the landscape had worn off and the strangeness was simply tiring. I longed for the simplicity of a stand of pine trees, the unchanging nature of rock.

Ambrose, too, was terribly homesick for the West. I was reading *Huckleberry Finn* to the boys, and he talked about lighting out for the territories. Whenever we passed a certain county road that headed northwest, disappearing over a small rise in the distance, he talked of putting his knife and compass in his backpack and heading off "down that road we've never been on before." One weekend he saw a movie that took place in the Alaskan wilds, and it made him so despondent that he came home and climbed the tree in front of our house and wouldn't come down for hours. His longing for the wilderness grew deeper and more painful the longer we stayed in Illinois.

The place I had turned to for solace—a dirt path in an unimproved section of the county park—had lost much of its regenerative power for me. The west end of the path led from the top of a glacial moraine down through a long sloping field, once farmland but now filled with thigh-high grasses. The path was bordered on one side by thorny thickets of Osage orange and hawthorn trees, planted as a windbreak in the days when the land was still under the plow. At the top of this mile-long section there was a dilapidated barn that looked, the boys said, like an illustration by the children's author Bill Peet. The lower half was brick, the upper half whitewashed boards topped by a sagging cupola with a rooster weathervane. Hidden in the tall grass was the foundation of another small building, perhaps a shed or grain bin. I often brought a book and sat on this cement wall in the shade of an apple tree, listening to the call of meadowlarks in the field.

The east section of the path, just a fifteen-minute walk from our house, led alongside a smaller meadow between two ponds. These were the best frog ponds in the area, where Ambrose and I spent one memorable afternoon catching frogs the size of his baby fingernail. We often saw deer here, and there were always many bright cardinals flashing through the dark woods.

The county had the regrettable idea of linking these two seldom-used sections of the park with a paved bike path three and a half miles long. They started by ripping out the mock orange bushes that had given off a scent of oranges and suntan lotion. A huge swath of asphalt, ten feet wide, was laid down, with mown strips on either side. Instead of rocketing across the path into the grass, grasshoppers now landed with an unpleasant clicking sound on the pavement. When the path had been simply a narrow dirt trail filled with puddles and muddy patches, I had to walk slowly and pay attention to where I was going. Now that the pavement was smooth and there was nothing to look at beneath my feet, I found myself walking faster. Like the bicyclists who sped past, I looked at the "view" of the fields around me as if from a great distance.

One day I drove to the west end of the path with my dog, intending to hike down to the cemetery and back up again to the barn. But instead of the barn to greet me with a flurry of doves flying out its cockeyed doors and broken cupola, there was nothing but the foundation and a pile of ashes and blackened bricks. I turned the car around and rushed out of the parking lot, crying so hard I could hardly see to drive. Unlike everything else in this respectable midwestern town, the barn wasn't neat, well ordered, and prosperous looking. It was just an old working barn that had seen better days, and a comforting reminder of the West, where cycles of boom and bust had littered the landscape with many such buildings. I felt robbed of one of the few things that had made life here bearable. I started to feel like I was sinking. I had to get out West.

The following week, we drove up to Chicago to spend the day. Emlyn and Richard were playing a virtual reality game at a high-tech arcade, and Ambrose and I, tired of the noisy mall, decided to wait for them in the parking lot across the street. I was doing a quick rundown of our finances in my head, trying to decide if we could afford a trip to the Northwest. Ambrose was sitting on the hood of the car, craning his neck to see the top of the tall buildings. Tears started trickling down his cheeks. "I hate Chicago," he said. "You can't even see the sky." He got into the car and curled up in the back seat, sobbing.

When he finally calmed down he told me, between hiccups, of his great longing to live in the wilderness, and of the vision he had of his future. He had it all planned out. He described the house he would build, the land and the creek running through it, even the rock that would be his favorite place to sit and think.

"I'll have to learn a lot of stuff first. How to build fences . . . take care of animals." He spoke softly, dreamily, as he described it all, "And when I'm lonely, I'll go to town for supplies and talk to people in the stores." He planned to make his living writing books and selling his paintings. There was a pause and then he wailed, "But it's all so far away! It'll be years before I'm old enough to do what I want to do."

I took him onto my lap and talked to him about things he could do in the meantime—skills he could learn, places he could explore. And I reminded him that he was almost ten and in a few years he would be old enough to help out at the ranch and he could spend summers there.

Richard and Emlyn came back, and I turned my attention to my older son as he chattered on about the cool game he had played—some kind of space battle simulation. On the three-hour drive home, Ambrose fell asleep, exhausted, his head on my lap. Richard listened to a tape of a new jazz piano piece he was working on. Emlyn read a book by flashlight in the front seat. In the dim light I looked down at Ambrose and stroked the still baby-soft skin on his cheek. It was time to go back to Idaho.

FOURTEEN

Going Home

Almost a year after we moved to Illinois, my family came back to the Northwest for a visit. We were so anxious to get out of the Midwest that we drove straight through to Wyoming in twenty-one hours. We camped overnight at Farewell Bend, where the Burns family had stayed on their way to Oregon Territory well over a hundred years ago. Instead of saying farewell to the Snake River, we were saying a heartfelt "hello again." Although it was nothing like the free-flowing Snake at the Burns Ranch, it was our first sight of a river we had come to love. The dry hills, slack water, and muddy banks were like the unappealing but familiar part of the Snake nearest Pullman, just below Lower Granite Dam.

After setting up our tent on grass as flat and green as a pool table, we went for a walk along the river. Killdeers screamed and ran at our approach, carrying on convincingly about their broken wings. We stopped where a stream meandered through a marsh and collected in a quiet pool before flowing on into the Snake. Emlyn was complaining that the mud kept sucking the thongs right off his feet, and Richard was making Felix Unger-like noises that meant he was developing a migraine. Our dog was running in and out of the water. Ambrose was looking for a good rock to throw. Suddenly, a wide furry head poked out of the water almost at my feet. As the noise and motion of my family continued around me, an otter looked me in the eye for a long, still moment. The otter slipped out of sight before anyone else saw it.

I told the ranger that evening and his face lit up with a mixture of joy and envy. "Yup, we've got river otter back in that marsh. You're really lucky. I've been here eighteen years, and I've only seen one once." I felt grateful to be acknowledged in such a way. I was indeed being welcomed back to the Snake River.

Old cowgirl crossing sign on the ranch house porch. Photo by the author.

Richard and Emlyn spent just a day at the Burns Ranch and then drove up to Pullman to visit friends and relatives. For both Emlyn and his dad, a little time at the ranch goes a long way. Richard was happy to see Bertie and catch up on news of all the local musicians. Emlyn had a good time checking out Liz's computer and Nintendo games. We had a peaceful afternoon at Ram's Head, where the boys waded in the river and skipped rocks, but by the end of the day Richard and Emlyn were ready to leave. Ambrose and I planned to stay over the weekend before heading up to the Palouse.

After Bertie ferried Richard and Emlyn across the river, we built a fire on the beach and played baseball. Ambrose hit the ball and sent it flying into the field of rocks at the bottom of the bluff. It bounced off a boulder, boomeranging wildly into Bertie, who was running to catch it. I took a turn at bat and Ambrose made spectacular flying leaps into the sand to keep the ball from going into the river. The western rim flared with the last of the evening light and then it was dark, and the sand was cool beneath our feet. The fire was a bright, mesmeric orange against the dark blue sky and the sharply outlined canyon rim. We roasted hot dogs over the fire, and Bertie told us the latest jokes she had heard from the banjo player in her bluegrass band.

I was perched on a piece of driftwood. Liz sat cross-legged in the sand. She didn't say much, just looked into the fire and laughed quietly at a joke now and then. Bertie put another chunk of wood on the fire. Ambrose pulled out his pocketknife and found a stick to whittle. The river, flowing just a few feet away from us, made sloshing, sucking sounds like a great animal feeding. The firelight on Liz's face brought out the faint marks of the old dog bite around her eye, like the ritual scarifications of some ancient tribe. She looked ageless and beautiful. I watched the sparks flying upward against the dark canyon walls and thought about how much I had missed this place and these people. The entire nine months we had been gone, they had had only one visitor: a photographer kayaking the river who asked permission to take a picture of the barn.

Our drive along the familiar Snake River Road to the ranch had seemed strangely lonely this time. I usually keep an eye out for bighorn sheep on the stretch between Asotin and the ranch. Sometimes I saw a half-dozen or more on the rimrocks across the river. But more often, I saw the Black Butte band on the Washington side. One day I stopped the car as I saw them coming up from the river. They crossed the road right in front of me, and I rolled down the window and said hello. They turned to look at me, flicking their ears. Then they continued on up the hill, casually walking right up to a fence before springing over it.

Now they were gone, and the road was empty without them. Over the winter, the entire Black Butte band had been wiped out, along with virtually all of the bighorns on the Washington side of the river. Just after Thanksgiving, the bighorns starting falling sick with bacterial pneumonia, and a massive effort was made to save them. Over seventy were net-gunned from a helicopter and flown down to southern Idaho for treatment. All but seven died. In all, over a hundred bighorn sheep died of a common bacterium that is carried by domestic sheep. The good news was that the bighorns on the Idaho side were not affected.

With this devastating epidemic, it looked like the end of sheep ranching in the canyon. Early in the century, the native bighorns had been hunted into extinction in the Snake River Canyon, so there had been no conflict between the needs of domestic and wild sheep during the years when huge sheep ranches ran thousands of domestic sheep on either side of the Snake River. But an ambitious transplantation program, started in 1971, eventually brought in two hundred bighorns from the Canadian Rockies.

It proved impossible to keep the two populations—wild and domestic—away from each other, especially since breeding rams are inclined to roam. For years, the U.S. Forest Service kept the situation under control by killing bighorns that wandered into domestic herds so they wouldn't carry foreign germs back to their own herds. For a while, it looked like the situation would eventually resolve itself as the number of domestic sheep in the canyon continued to dwindle. Still, there were a few landowners, reluctant to abandon ranching altogether, who would put a band or two of sheep out there every few years just so they could hold onto their sheep permit.

Even Dooley, who hated to see the sheep go, has to admit that when push comes to shove, the advocates for the bighorns are going to win. At this point, bighorn hunting is bringing more money into the local economy than sheep ranching. To me, it just seemed like a double whammy: dying bighorns, departing domestic sheep. The canyon is changing, no doubt about it. The first thing I asked Liz, when we arrived at the ranch, was, "Did they log up by Percival's house?"

Liz said, "Yeah. I can't stand to go up there anymore. It's too sad."

Throughout the rest of our visit I questioned anyone who might know anything about the logging. I wanted to know exactly what had taken place. It was like asking to hear the grisly details of a loved one's death, but it was somehow consoling that I was not the only one feeling a sense of outrage and loss.

The field officer for the Nature Conservancy said, "They really hit it hard." Even over the phone, I could clearly hear the sadness in his voice. "We negotiated to buy that property for years. There were huge old ponderosa pine there that they didn't cut the first time, back in the fifties. Now we've lost over ninety percent of the old-growth ponderosa on the mountain."

Dooley, with fifty years of selective logging on his own land, was critical of the shoddy practices of the wildcat logging operation. He said disgustedly, "They cut down to six- to eight-inch-around trees. They didn't leave none in the production state. Moved a million board feet of lumber, and damned if they didn't lose money doing it. They cut some high-quality pine and got about a thousand dollars per thousand board feet. But then they cut on the hillside. It's so arid up there, it's all pitchy, knotty pine. Went for pulp. They only got a goddamn hundred dollars per thousand board feet for it, when it cost 'em three hundred dollars to airlift it outta there!"

The Fish and Game representative was even more blunt. "They lost their butt," he said. "They took it all—everything that was merchandisable. But it was a money-losing proposition from the word go."

The hillsides were so steep that the only way to get the trees out was by helicopter. Although it's always more expensive to use helicopters than logging trucks, the cost can usually be minimized by flying the timber downhill to a landing site and trucking it to a mill. But on Craig Mountain, they had to airlift the trees eighteen hundred feet up and over the ridge to a landing site on Fish and Game land. It was slow going, flying uphill with a heavy load, and it consumed a lot of fuel.

The Fish and Game department spokesman went on. "Using helicopters meant no new roads, so it was the best possible way to go for the stream, if it had to be logged. We're very concerned about Thornbush Creek. We kept a seventy-five-foot buffer on either side of the creek."

Dooley scoffed at that number. "They cut down to within twenty feet of the creek in some places."

Liz agreed. "They cut down some of the biggest trees along the creek, and left trees only this big around"—she made a circle about the size of a salad plate—"that aren't nearly big enough to shade the creek."

"Well," the Fish and Game guy admitted, "it was a process of give and take. We said, 'You can take this big tree by the creek, if you leave these three medium-size ones over here.'"

Later, Liz protested, "That's not the way it's supposed to work." She sounded close to tears. None of the other problems she faces on the ranch bring the sense of helplessness she feels in dealing with the Fish and Game Department.

Over the winter, the Burnses' fears about the creek were justified. Rain and snowmelt coursed down the denuded hillsides, flooding the creek so heavily that it jumped its banks and carved out a new course down a section of the dirt road. With the road washed out, the Burns couldn't get to their summer pasture or drive to town via Waha.

Although the sound of the helicopters has long since ceased, the silence around Percival's house has not returned. The hushed stillness that had been the most remarkable feature of the valley is gone. Without the trees to serve as a windbreak, Dooley says, the wind just roars down the mountain.

Dooley was staying down at his house in Asotin, so I drove to town to see him. I left Ambrose at the ranch; he and Bertie were busy "slaughtering lawn chairs" with arrows and bows they had made from syringa branches.

On the south side of Asotin, I headed up the highway that leads to the Anatone Prairie and on to the town of Enterprise, Oregon. I turned at the sign that says, "Next Services 77 Miles" and pulled into Dooley's steep driveway. It was a gray, chilly day, and he turned on the stove burners in the kitchen to warm up the room. The old rancher had suffered a slight stroke a few months earlier, and as he moved about the kitchen, I could see that he was not as steady on his feet as he used to be. He turned up his hearing aid and settled down to talk.

I enjoy talking to Dooley, although he is known for going off on long tangents about geology or his unique theories about how the world economy operates. He asked me about the weather in the Midwest and then said, "I haven't been very many places, but I've been to Hawaii," and he launched into a detailed explanation of the structure of volcanoes. I had often admired his knowledge of the geology of the Snake River Canyon, although it is not always in accord with the views of current scientists. I envied a man who understood enough of the mechanics of how the world works to fill in the blanks and create his own unique view of life. I had wondered, over the years, what he would have made of himself with more education and a wider experience of the world.

"You should have been a geologist."

"What for?" He laughed at the thought that he would want to be one of those damn fool scientists stuck in a lab somewhere instead of out on the river. He sounded like a man with few regrets about the life he had chosen.

Dooley had some old photos that he wanted to show me, so we moved into the combined living room/bedroom, with its huge picture window overlooking the river far below. This was where Liz and Gina had lived during the winter when Liz was in school and the road was too bad to go back and forth from Ram's Head to town every day. I had often thought that it was no wonder Liz and her mother hadn't got along; they must have been on top of each other in this tiny place. The whole house was only twenty-eight by thirty feet, with the addition of a camping trailer out back that served as Liz's bedroom.

In front of the piano sat an outboard motor on a piece of plywood. There were several framed photos on the wall, including two of Dooley with his arm around the neck of the tame elk that used to hang around the ranch. The elk showed up one day in the horse pasture. He had been raised by hand by someone up on the mountain and then turned loose. He was unafraid of people, and he soon developed an attachment to Bertie. Liz called me one day and said, "You oughta come down and see Bertie's new boyfriend." When we arrived at the ranch, the young bull was lying down comfortably in the shade next to the house, with the horses standing nearby. He nuzzled Bertie as she scratched between his antlers. Ambrose moved closer, and the elk's nostrils flared as he sniffed the seven-year-old's hair. The next time I saw the elk, months later, he was missing an antler. He must have been scared by a coyote or by people on the beach and slammed into a tree as he was running away. Having people teasing a half-tame elk was just an accident waiting to happen, especially in the fall when he went into rut. The Fish and Game Department arranged for what Liz called their "resident forklift" to be transported to a lab in southern Idaho.

Next to the pictures of the elk was a photo of Dooley on the deck of the *Caroline*. Refurbishing the *Caroline* was a project fraught with difficulties, and every time I see him, Dooley has some new story about his dual nemeses: vandals and government bureaucrats. Vandals broke in repeatedly, throwing the furniture overboard, pouring paint everywhere, and trying—stupidly— to set the steel boat afire. The Army Corps of Engineers, who managed the waterway, and the city of Asotin, which managed the city park where the *Caroline* was moored, gave him endless grief.

Dooley and his partner, Lon, were required to install railings, remove leaded paint, repair stairways, and take countless other steps to make the boat safe enough for public gatherings. In an attempt to deter vandals, they hired someone to live on board as caretaker, but the city squawked. The park was "day use only," they said. Dooley threw up his hands. "Nobody's ever explained to me why all the ships in the world have people living on 'em except the *Caroline*!"

Dooley is nothing if not stubborn, and he has doggedly carried on his fight to save the *Caroline*. But recently he suffered a blow that, for the first time, made me doubt the future of the old paddle wheeler. Lon, whom I remembered as the boisterously cheerful man behind the camera at the Burns family reunion, had drowned in the Snake. He fell overboard while returning home after the annual lighted boat parade on Christmas Eve. His body wasn't found until four months later, when it appeared far downriver below one of the dams. Dooley had taken the loss very hard.

When I asked him about the *Caroline*, he didn't rail against the latest damn fool stumbling block the government had put in his way. Perhaps recognizing that he couldn't go it alone on such a vast project, he said, "I'm thinking about trying to get the local historical society to take it on as a project."

I hated to put an end to our conversation, so it was late when I left Dooley's house and headed south along the river. Dusk was falling quickly and I worried about getting across the river before dark, but I knew better than to speed on this winding road. There were always big chunks of fallen rock to avoid, and the deer were out at this time of evening. I wished I had brought a flashlight. I would have to get across the minefield of shin-busting rocks down to the water in the dark.

By the time I pulled into the garage and padlocked the metal grate across the opening, the canyon was dark, but I was relieved that the sky held enough light for me to make my way to the boat. The river had dropped during the day, and the rowboat was stuck high on the rocks. I felt around for the clasp to the chain and wrestled the boat down to the water, the metal hull scraping and banging. Once on the water I relaxed. I was no longer in a hurry. The river was low and not moving fast. All I had to do now was row strong and straight. I could barely see the opposite shore, so I concentrated on the notch in the rim directly above the garage, outlined faintly against the western sky. When I was more than halfway across, I rested on the oars for a moment and looked over my shoulder. The beach had disappeared into darkness, but there was a light bobbing down the hill. Liz called to me, "Don't stop. Keep rowing." I pulled on the oars until the boat slid into the cone of light shining on the water and grated to a stop in the sand.

Later that night we went outside to look for the comet Hyakutake. It wasn't hard to find. It blazed at the corner of the Big Dipper, brilliant against the black sky. I called Emlyn and Richard at our friend's house in Pullman, but it was foggy up on the Palouse, and they couldn't see the comet. I pulled

Ambrose out of bed, blanket and all, and took him outside. He stood in the yard for a moment, swaying sleepily until he got a good look at the comet. "I see it," he said, teeth chattering. "Can I go back to bed now?"

Liz and I couldn't bring ourselves to go inside, although we were shivering too. Looking into the black sky punched with brilliant five-point stars, I thought of the dots and whorls pecked into the rocks at Ram's Head. We speculated whether there had been anyone in the canyon to see the comet the last time it had come around. It appeared eight thousand years ago; the oldest petroglyphs at Ram's Head are almost that old.

In the dark, we talked about the National Park Service's attempts to get a lease on Ram's Head. The Nez Perce National Historic Park has sites scattered around Idaho and Montana that illustrate important events or places in Nez Perce history. Some of the sites are on federal land and other locations are leased from private owners. The Park Service has Ram's Head listed as a "high priority" site, citing the need to protect the petroglyphs from vandalism and theft.

Protection for the petroglyphs on the Washington side of Ram's Head seems justified because it has already suffered a great deal of damage. Over one-quarter of the pictures have been vandalized—painted over, scratched away, or obliterated by bullet holes. And that's not counting those that were destroyed when they blasted the roadway in the first place. Because the site is so easily accessible, it also makes sense to make it a small roadside addition to the park system. Apparently, the new owners of the land (who bought it from the original homesteading family a few years ago) agreed: they donated the site to the Park Service. Development plans include a walkway down to the beach, interpretive signs, picnic tables, and a restroom.

In addition, the county has on its books a project to move the road higher up the bluff. They claim the reasons are twofold: too many people have gone into the river at the sharp curve at Ram's Head, and the petroglyphs would be easier to protect if the road were further away from the rocks. Liz is unhappy about the road project, especially since it would be up where the cliff was more unstable, possibly leading to landslides or rock falls. Lewiston has seen the effects of cutting these steep hills too drastically. A road cut made to widen a road just south of town caused a huge creeping landslide that closed the road for months. Dooley thinks the plan to move the road at Ram's Head is just "another goddamn dumb government deal" because the proposed route would disturb an ancient village site.

My objections were less practical. Although we did not talk about it, I'm sure Liz would agree that trying to make Ram's Head safer and more accessible is a mistake. Ram's Head is not just a pretty place to have a picnic, and it should not be reduced to that. Its wildness is its power, its danger its attraction. I know that Dooley would disagree. The dams upriver have weakened Ram's Head's powerful eddies, and Dooley says, "Those swirlpools—I, for one, don't miss 'em one damned bit." But something of the river has been lost at Ram's Head, although the stretch is now less dangerous for inexperienced boaters. Does the road, too, need to be made safer for people with poor judgment who exceed the speed limit on the winding gravel or drive back to town drunk after a day of fishing? How far will we go to protect ourselves?

If deciding the fate of the Washington side of Ram's Head is difficult, considering what to do with the Idaho side, which has a much larger collection of petroglyphs, is even more difficult. On the positive side, the site is accessible only by boat, which cuts down on the number of visitors. The majority of people who view the site are with a tour group; the presence of the guide and the fact that he keeps them moving from one panel to another may serve as a curb on vandalism.

Caring for the petroglyphs has always weighed heavily on Liz, who feels it is her personal responsibility. Trying to keep an eye on things at Ram's Head has kept her tied down to the ranch, especially on holiday weekends when there is increased boat traffic on the river.

Liz sighed, "Maybe they're right. Maybe they could protect it better. I can't be down there all the time."

But she is concerned that limits would be put on what activities are allowed there. Would she be able to make changes or improvements to the buildings there? What about the cows that graze on the bar in winter? Would she be able to build a fire in the wide circle of stones in front of Ram's Head Rock? Could Bertie still beat her great one-sided drum on moonlit nights?

"It's a holy place, not a tourist attraction," she said.

The Park Service proposes to station a ranger at Ram's Head to provide constant supervision. They've even suggested ways they could improve the cabin to make it more livable. I imagined with distaste a ranger hanging his hat on the nail behind the door. An official presence would ruin the sense of solitude that I come to Ram's Head to find.

In other areas of the Northwest and Canada, petroglyph sites have been fenced off and raised walkways built to keep tourists at a distance. The thought of having to peer through a chain link fence to see the small red figures made me feel sick and angry.

Some sites on federal lands, particularly in the Southwest, are kept secret from the public entirely. In a recent report on the proposed road project, archaeologists stated that an "avoidance protection strategy" wouldn't work at Ram's Head because everyone in the county knows where it is.

"Making it part of the park system will just bring in more people," Liz said, "And more people means more chances of vandalism. But what scares me to death is something I've had several people tell me lately." She took a deep breath, reluctant even to say it. " 'I'd be doing you a favor to come in here and blow up these rocks. Otherwise the government is gonna take your land.' "

"What did you say to them?" I asked, imagining the whole weight of Liz's wrath coming down upon some poor idiot who thought he was on her side.

"I gave 'em hell! I read them the riot act!" she sputtered. "It'd be such a tragedy if that happened—like this pretty little lake up by Winchester. The owner took a bulldozer to it so it wouldn't be declared a wetland."

The words of Dooley's old friend Dorrie McAlpine came back to me. I had run into her a few days earlier at the Asotin Museum. About Ram's Head she had said, "The Burns family have protected it as well as anybody could. It's *theirs*—it belongs to them. The National Park will lock it up— keep everybody away. They just want a toehold on Ram's Head, and then they'll try to get the rest of the ranch. The Burnses are practically the only ones left on Craig Mountain now. The government's got them surrounded and they're putting the squeeze on them." When she rose to leave, her last few words sent a chill through me. "It might not come in Dooley's lifetime, but the time will come, not too many years from now, when Liz is going to have to fight like hell to keep her land."

The next morning, Liz and I went out to the corrals in the frosty morning to feed the calves. As we pitched hay into the troughs, I asked her about the disturbing changes coming to the canyon. I had always assumed that Liz's tie to this land was rock solid. In the ever-changing West, one hundred years is a long time. Surely, I had thought, if the Burnses had been here for over a century, they would continue to be here into the foreseeable future. That was before I learned that they had weathered a thirty-year running battle over the Asotin Dam project. By the time plans for the dam were abandoned—about the time I met the Burnses—Liz was approaching middle age. Her concern about keeping the land had shifted to her own ability to carry on the heavy work as she grew older.

"It gets harder every year," she had told me as we watched the calves eat. "The winters are always hard when the roads are real bad. We're stuck out here for weeks at a time with no one but each other to look at."

I thought about father and daughter living in separate houses, a stone's throw from each other, and Dooley's words, "No ways could I live in the same house with her. Lizzie's got to have everything done a certain way. She goes about everything backwards." In recent years, he's taken to spending much of the winter in his house in Asotin.

Liz continued. "I move slower in the morning than I used to. The cold is getting to me more. I'd like to get off the ranch and travel, but it's getting harder and harder to find a chunk of time when I can leave things here."

"You're supposed to have an eighteen-year-old son to take things over about now so you can start easing up," I teased.

She paused at the garden gate. As her hands worked the jury-rigged latch, she looked off toward the river. "It probably wouldn't have turned out that way. He might have gone off and not wanted to ranch. But I think I would have been a good mother. I'd like to have had the chance."

"So what do you do?"

"Just keep on until my body gives out. Then I'll probably have to lease the ranch and move to town. Maybe I could lease just part of it and still live here. I don't see how I can live in town again after all these years."

But in the years since that conversation took place, more ranches have folded, and more ranchers have retired or sold out. There aren't many people left who would be interested in running cattle in the canyon. Now, looking up at the snowy bulk of Craig Mountain looming over the ranch, Liz can't shake the feeling that the government is crouched, vulturelike, up there, "just waiting to starve us out."

The last day on the ranch, we walked through the orchard, past the schoolhouse and the remains of Dooley's garden, out to the pasture where Belle was buried. Usually, when a large animal dies on the ranch, it's simply dragged to some out-of-the-way place where the smell won't be too bad. But Belle had been Liz's favorite horse for many years. She had scooped out a hole with the tractor and buried the horse deep enough so the coyotes wouldn't get at her. We sat next to the long mound in the grass and watched Ambrose swing himself into Dixie's saddle—he didn't need a leg up anymore—and ride across the pasture to the line of apricot trees at the far end. He wheeled the horse around and urged her on until her mane flew.

Earlier that day he had found an arrowhead at the water's edge where a recent rain had crumbled a sandy ridge along the beach. After examining the small, salmon-colored arrowhead to make sure it was real, Ambrose had crowed with delight at his good fortune. Now he whooped again as he raced across the pasture.

"You don't have to worry about Ambrose," Liz said.

I thought that she was just calming my fear that he was riding too fast, but as she continued I realized that she was talking about his future.

"Look at him," she said, as he rocketed past, his face alight with joy. "A few years in the city aren't going to hurt him. He'll always be a nature kid."

I leaned back in the grass and looked up at the rimrocks outlined against the pale spring sky. I felt a great sense of relief that time and distance had not broken our connection to this place. My love for the canyon was a constant in my life, giving me a sense of hope about our uncertain future.

I scrambled out of the truck when we got to Ram's Head, eager to climb the rock again and look out over the river. I sat on my heels, hugging my knees and humming to myself. My eyes fell on the figures in red carrying unfathomable messages in their outstretched hands. I thought of the time I had stopped at a table full of Indian crafts at a Nez Perce powwow. Among the beaded purses and belt buckles was an odd-shaped curio. I thought it was some kind of ceremonial object, an armband perhaps. "Excuse me," I said respectfully, "but what is this used for?" The woman behind the table said, her voice bored and scornful, "It's for hanging on your rearview mirror." And I saw it was a cheap little beaded headdress, the Indian equivalent of a pair of fuzzy dice. I felt foolish—here I was again grasping for some hidden meaning.

At Ram's Head I was forever trying to catch hold of something and take it home with me: the Sputnik button from the old shack, the key to the cabin, the true meaning of the petroglyphs. I suddenly felt profoundly ungrateful. Enough of this taking things. It's time to leave something. For the first time in my life, I wished that I smoked; tobacco is the traditional gift to leave for the spirits. Feeling around in my pockets, my fingers touched the frayed end of a roll of Lifesavers. I poked one of the candies into a crevice in the rock. I hoped the grandmothers liked butter rum.

We stopped by Dooley's house to say good-bye. At the door, Dooley greeted us and said to Ambrose, "I see you've been out riding the damn horses." Ambrose grinned and touched the long double scratch he had gotten on his cheek when Dixie ran under a low-hanging branch. Ambrose wandered outside while Dooley and I talked. I asked how he planned to deal with the Park Service.

"I told them there's no damage being done here now. All them tourists— probably several thousand people a year stop there and go up there and look at them Indian writings. They don't hurt a damn thing. The main damage is from when they did construction of the road over on the Washington side, and that had to be done. But if you went to squabbling with them, I'm afraid you'd get something a-going."

"Do you think they'll stop bugging you about it if you just keep saying no?"

He laughed with a sense of acceptance, of knowing the nature of the beast. "There's no such thing as them stoppin' bugging. They're that way. They'll be that way as long as the earth is here."

The changes and losses to their way of life, which Liz suffers so bitterly, Dooley accepts with good-humored resignation. I didn't know how much of that has to do with his age. Perhaps he knows he won't be around to see the bitter end that everyone in the canyon knows is coming. Next to the photo of the elk on Dooley's wall is a framed poem. The cowboy poet who had been at the reunion wrote the poem "on behalf of Dooley," and it expresses the sentiments of many local residents. A lament for the passing of the ranching culture, it ends:

One hundred years and I stand alone.
The last of a natural breed.
Created by sweat and love and blood.
The child of a homestead deed.

Dooley may indeed be the last of his kind, but it is Liz who is the last of the Burnses. Without an heir, the Burns Ranch may go the way of many other ranches in the canyon.

I asked Dooley what he saw down the road, and he said, "I can't see there being ranches in the canyon much longer. The government will get all of it eventually. People don't want to live a hard life anymore. It's too easy to live in the city."

I noticed that Ambrose had slipped into the room.

Dooley was talking about some place in Asia he had read about recently, where young children are housed and educated at their workplace, learning how to be good employees. "They live in these dormitory rooms that are just tiny—it's like a honeycomb. They got no freedom, no place to roam. They don't develop into real people who have a chance to *do* anything. They're just programmed to work."

His words echoed what Liz had said just the day before. I had asked Liz what she would like to see happen to the ranch twenty-five or fifty years from now. She had leaned on the fence and watched the calves feeding. They were fat and healthy, and they gave off a warmth that felt like life itself rising into the cold air. "I'd like to see a family living here again. The way kids are raised these days, they don't have a chance to ever be real human beings."

I felt a sense of despair that the world both father and daughter described was the one my sons would inherit. A world with plenty of worker bees but not many true human beings.

"You're lucky to have experienced that kind of freedom before it disappeared," I said.

Out of the corner of my eye I saw Ambrose leaning in the doorway, listening intently.

"It's still there," Dooley said, "You just gotta look for it."

FIFTEEN

This Perfect Place

We left California when our children were ages three and one. Neither of them remember having lived there. I tell my sons that Santa Cruz is a cool place to have been born and they should be proud of it. They have grown up as small-town kids in the "lentil capital of the world." I wonder about the impact on them of being raised in the margins, in a state described in East Coast news reports as "remote," in a town recognized only for its college football team. When we travel, the boys have to explain that they live in Washington State, not Washington DC, and that no, it doesn't rain a lot here on the east side of the state.

When I say I'm from L.A., it needs no explanation. Even here in Pullman, people recognize the name Redondo Beach. They invariably say, "That's a really nice area!" with the question in their eyes, "Why did you leave? And what are you doing on the Palouse, of all places?"

I felt a wrenching sense of loss and shame upon leaving California. Not only was I leaving my much-loved family and friends, but I was refusing a wonderful gift our great-grandparents had given us: a precarious hold on that beautiful corner of the universe. It seemed like an act of supreme ingratitude.

Leaving the beach is like losing one's religion. It's a failure of sorts. We've turned our backs on what millions of people have agreed is a reward worth having, or at least striving for.

Do I miss it? For years after leaving the ocean, I used to explore that spot in my mind tentatively, like a kid touching his tongue to the new hole in his gums where a tooth used to be. Years later, the answer is still, "Nope, doesn't hurt." I enjoy knowing there is land, good solid land, surrounding me on all four sides. The climate agrees with me. I remind myself that my ancestors all came from the northern latitudes; the sojourn in California was just a hundred-year detour.

In the thirteen years I've been gone from California, the twenty-four years I've been gone from L.A., my memories have grown dry and bleached like the ashlike coating on our legs after a day at the beach. (This is what I imagine old age will be: a dry slipping of memory against memory, ice against ice, like Smokey heard in her old-age winters as she woke to the tide shoving frozen shards onto the beach.)

When I visit Los Angeles now, I'm amazed at how green it is. In my memory it is a desert. I understand how my ancestors coming from the Midwest must have felt upon arriving in California and encountering the sun, the warmth, the bright colors. Everything seems as vivid and fast-moving as the flocks of escaped parrots that swoop through the neighborhoods of the beach cities.

My children do not understand what we found so awful about L.A. I can't explain to them my teenage feeling of claustrophia, of being stuck, of how big a state California is, and how hard it was to leave. It's a sad thought that in my search for a home, I have made myself and my children strangers to their birthplace and mine.

The truth is that I could not have stayed in Redondo even if I had wanted to. With the mass of humanity pushing its way toward the shore, we were all eventually nudged right off the map. My siblings, my childhood friends, their brothers and sisters and parents have all had to move to outlying areas of the city more affordable for families and retirees. One of my sisters was able to remain just long enough for her son to become the third generation of Freemans to graduate from Redondo Union High School.

My parents thought that one of the benefits of living in the city was that their children would be able to live close by when they were grown; unlike young people in economically depressed small towns, we would not have to go away to find good jobs. In the 1960s, when housing was affordable and the beaches swarmed with children, my parents could never have imagined that we would leave the South Bay someday, searching not for work but for a roomy house with a yard in a peaceful neighborhood.

The curve of the bay, the salt breeze, and the sound of the waves provided some continuity of experience for five generations of my family in a fast-changing South Bay. Now my siblings are scattered from one end of L.A. to the other—from Ventura County in the north to Orange County to the south—as well as from Idaho to Texas. None of us is within sight or sound of the ocean.

Ex-Californians like us are spread throughout the country, and there is no shortage of them in the inland Northwest. When I meet someone from California we share an unspoken agreement not to call too much attention to this fact, as well as a rueful acknowledgment of the questionable success of our efforts to "pass" in local society. There is also a recognition among members of this diaspora that we will never finish the transition from Californian to Washingtonian or Idahoan; we are labeled ex-Californians, a term we will never shed. We'll be stuck forever in mid-molt. Part of the reason we stay so immutably Californian is because that is the only thing we are allowed to be.

Native northwesterners are proud of their strong prejudice against Californians. Soon after we moved to the Northwest, an acquaintance found out I was from California—information I rarely volunteered, but readily admitted if asked. "It's OK," she said magnanimously, "Some of my best friends are Californians."

I cannot expect the natives of this area to accept me or to approve of the choices I made that brought me here. I have to look not to the local pioneers but to the pioneers of my own family. Uncle Herman thought my teenage dream of moving to Colorado was a bunch of nonsense. "People are the same everywhere you go," he said from his bed in the nursing home—just about the last words he spoke to me. But he would have understood my wish for a large yard full of fruit trees and the pleasure I get now from each plum, apple, and cherry. Smokey knew the magnetic pull of the North and would have shared my delight in the long, long evenings of summer. Grace, who was yanked out of college and put back in the kitchen, would have recognized the lure of higher education that eventually led us to the Midwest for three long years. My ancestors would have understood why I left California.

If we honor the westering movement of our ancestors, we should accept the same drive that moves more recent emigrants. Our reasons for coming to the Northwest were no different from those that led the Howard, Inglis, Rhodes, and Johnston families to leave their comfortable homes in the Midwest: a belief that we could do better for ourselves in a new place, a desire for a healthier environment, and a longing to be reunited with family members who had gone on ahead (my younger sister had moved to the Palouse three years before us). We were simply continuing on west of California. Sometimes to find the true West, you have to go in a different direction entirely.

My grandmother found her West in Alaska, a place where she could cast off the remains of the privileged upbringing she found so stifling and remake herself as a sourdough named Smokey. It was easier for her, in some ways, to work herself into a new landscape. I had trouble getting to know the Northwest, not because it was a different place than I knew, but because I had never known a real place before. Smokey, at least, had some experience relating to a natural environment: she could compare fishing from an open boat with surf fishing, catching euluchon with catching grunion, cooking mussels to cooking abalone. She had the sense of being from a particular region with its unique climate, plants, and animals. I had no such sense. If I had had that connection to the land, perhaps I could have weathered the changes in California and stayed.

Instead, I have skimmed lightly—lived in six different states, on both coasts and in between. I think of Liz walking the same daily routes over the years, wearing those trails through the meadows, circling deeper and deeper, spiraling into the core. I can only understand where I am in relation to where I've lived before, triangulating my position from Maryland on the East Coast and California on the West. For one whose family has lived here for generations, this kind of locating device is unnecessary, but for me, it places a pin firmly in my personal map. It is my own "You are here."

I left L.A. with a vision of the mountains. I never made it to the Rockies or that little place in the country at the end of a dirt road, but I did land in this small town, just as my forefathers settled in Iola and Florence (those towns like the names of sweethearts); it is my Osage Mission, my Crab Orchard. My great-grandparents and I share things I could not have known before: the sound of the noon whistle, the gossip and grudges of small-town society, the smell of wheat from the fields during harvest; the warm embrace of the community at holiday celebrations in the park.

In trying to anchor myself in space, I've managed to fix my position in time as well. My sense of family has expanded to include people long dead and children not yet born. Now that I am on a first-name basis with many generations of my ancestors, I am less lonely for my dead parents; they have joined a group of family members who are no longer strangers to me.

My children and their cousins are the seventh generation of our family to have lived in California. Some Native Americans believe that a decision will be made wisely if its impact on the next seven generations is considered. By this reckoning, perhaps my ancestors' decision to come to California and help build the boomtown of Los Angeles was the wrong one: their efforts to make a better home for themselves and their children led to a city too crowded, too violent, too expensive for the seventh generation. But the people of Southern California could not have predicted that the life they created would be one sought after by people all over the world.

A tow truck driver hauling our car up the Lewiston grade after a breakdown once told me, "I went to L.A. one time to visit my cousin. Back in the fifties. Came up over that pass and saw all those lights spread out and I asked myself, 'How am I gonna find one little porch light out of all those?'"

I never saw Los Angeles as individual porch lights—only a country person would view it that way. When we came up over Cajon Pass, the never-ending lights said I lived in a city that mattered. Small towns may wither and die, struggle along by themselves, and no one much cares. If something happens to L.A., the whole world notices. Californians grow up knowing they count.

Over the course of ten years, with few words and fewer direct glances, relying mostly on her actions, Liz told me this: You don't count as much as you think you do. None of us do. Practice humility. Be willing to look foolish once in a while. Get those feet under you, show some gumption, stand up to the rock. Stop thinking in the abstract about the environment, the economy, politics. Start seeing individual porch lights. Care about *these* animals, *these* native plants, *these* people, this perfect place.